SERIES PREFACE

Biomaterials science is concerned with surgical implants and medical devices and their interaction with the tissues they contact. Their study, therefore, includes not only the properties of the materials from which they are made, but also those of the tissues which will accept them. Metals, ceramics, and macromolecules are the artifacts. Bone tendons, skin, nerves, and muscles are among the tissues studied. Prosthetic materials, implants, dental materials, dressings, extra corporeal devices, encapsulants, and orthoses are included among the applications.

It is not only the materials *per se* which interest the biomaterials scientist, but also the interactions in vivo, because it is at the interface between implant and tissues that the success of a procedure will be decided. This approach has led to the concept of a more aggressive role for biomaterials in the actual treatment of disease. Macromolecular drug delivery systems are receiving considerable attention, especially those with the capacity for targeting specific sites in the body. Sensing and control of body processes is a logical extension of this. There is much to be done before these newer developments become established.

The science of biomaterials has grown and developed over the last few years to become an accepted discipline of study. It is opportune, therefore, to systematize the study of biomaterials in order to improve their application in medical science, since that is the end point of all studies. That is the aim of this series of books on *Structure-Property Relationships in Biomaterials*. Knowledge of structure and the influence on properties is fundamental to any materials science study; it is a more complex problem to obtain the knowledge from tissue materials, as the living organism has a great capacity for change and adaptation in response to a stimulus. The stimulus may be chemical, electrical, or mechanical. The biomaterials scientist endeavors to identify and to use these stimuli and responses to improve the in vivo acceptability of the materials.

Many institutions and agencies have promoted the science of biomaterials. Societies now exist for this purpose. The Biological Engineering Society (U.K.) founded in 1960 formed a Biomaterials Group in 1974. In the same year the Society for Biomaterials was founded in the U.S. The European Society for Biomaterials (1976) was followed by Canadian and Japanese Societies (1979). All societies play a major role in disseminating knowledge through conferences and publications.

This series is complementary to these society activities. It is hoped that it will not only provide a basis of knowledge, but also its own stimulus for further progress. The series is inevitably selective. In part this is due to the editors' choice, in part to the availability of authors. The editors wish to thank those who fulfilled their agreements. Without them this series would not have been possible.

G. W. Hastings
Series-Editor-in-Chief

VOLUME PREFACE

The understanding of the in vivo performance of synthetic materials is largely dependent upon a profound knowledge of the properties of the materials in question. Analogous to materials science in its broadest sense, the basis for biomaterials science is formed by microstructural theory. It is, therefore, that in this series on structure-property relationships in biomaterials a substantial part is devoted to the analysis of the basic properties of the various synthetic biomaterials. In addition, the effect of microstructural aspects on properties is considered at great length.

The study of metallic and ceramic biomaterials is intimately interlinked because the microstructural aspects and the research methodologies are founded on the same basis. This is demonstrated first in the chapter of Dr. Heimke, who discusses the structure of metals and ceramics, and second, in the chapter of Drs. Arkens and Ducheyne who analyze some of the more recent surface analytical techniques. Those techniques have been equally employed for metals and ceramics. Two subsequent chapters deal with the influence actual manufacturing has on both the microstructure and the properties of either metals (authored by Dr. Pilliar) or ceramics (written by Dr. Doremus).

The second volume of *Metal and Ceramic Biomaterials* builds upon the foundation laid in the first volume by analyzing various properties and by discussing these in light of the microstructural theory outlined in the first volume. Strength related behavior is treated first. Dr. Semlitsch surveys the mechanical properties of implant metals used for artificial hip joints. Drs. Soltesz and Richter describe the mechanical behavior of various bioceramic materials. A third chapter relating mechanical behavior to microstructural detail deals with the shape memory alloys which have considerable potential as biomaterials. In addition to the description of strength aspects, Dr. Kousbroek summarizes the various biocompatibility studies on these metals. These studies are, quite naturally, related to the surface properties which are presented in the second part of Volume II. Drs. Lycett and Hughes survey the corrosion behavior of metals. Drs. Van Raemdonck, Ducheyne, and De Meester analyze the bioreactivity of a typical class of ceramics, viz. calcium phosphate. An effort is made to relate the bioreactivity to the microstructural chemical and physical detail. Drs. Dumbleton and Higham analyze the field of surface coatings and point to the great potential for the field of biomaterials of many existing technologies.

The editors have enjoyed collaborating with the contributors to these volumes. We owe great appreciation and gratitude to those who eventually made it possible to have these books produced. Without each single contributor it would have been hard, if not impossible to present the information on some of the more important properties of biomaterials. Due thanks are also expressed to Rita De Laet who diligently took care of the secretarial work.

Last but not least, thanks go to our families. We may think that without their delightful distraction we could have finished these books sooner. However, they rightfully claim that without them we would never have had the perseverance needed to edit this series.

Paul Ducheyne
Garth Hastings

Metal and Ceramic Biomaterials

Volume II
Strength and Surface

Editors

Paul Ducheyne, Ph.D.
Associate Professor of
Biomedical Engineering
Associate Professor of
Orthopaedic Surgery Research
University of Pennsylvania
Philadelphia, Pennsylvania

Garth W. Hastings, Ph.D., D.Sc., C.Chem., F.R.S.C., F.R.R.I.
Acting Head
Biomedical Engineering Unit
North Staffordshire Polytechnic
Honorary Scientific Officer
North Staffordshire Health Authority
Medical Institute
Hartshill, Stoke-on-Trent
England

CRC Series in Structure-Property Relationship of Biomaterials
Series Editors-in-Chief
Garth W. Hastings and Paul Ducheyne

CRC Press, Inc.
Boca Raton, Florida

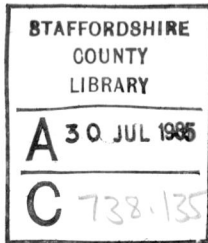

Library of Congress Cataloging in Publication
Main entry under title:

Metal and ceramic biomaterials.

(CRC series in structure-property relationship of biomaterials)
Bibliography: p.
Includes index.
Contents: v. 1. Structure — v. 2. Strength and
surface related behavior.
1. Ceramics in medicine.   2. Metals in surgery.
I. Ducheyne, Paul.   II. Hastings, Garth W.   III. Series.
R857.C4M47   1984   610'.28   83-15018
ISBN 0-8493-6261-X (v. 1)
ISBN 0-8493-6262-8 (v. 2)

This book represents information obtained from authentic and highly regarded sources. Reprinted material is quoted with permission, and sources are indicated. A wide variety of references are listed. Every reasonable effort has been made to give reliable data and information, but the author and the publisher cannot assume responsibility for the validity of all materials or for the consequences of their use.

Direct all inquiries to CRC Press, Inc., 2000 Corporate Blvd., N.W., Boca Raton, Florida, 33431.

© 1984 by CRC Press, Inc.

International Standard Book Number 0-8493-6261-X (v. 1)
International Standard Book Number 0-8493-6262-8 (v. 2)

Library of Congress Card Number 83-15018
Printed in the United States

THE EDITORS

Garth W. Hastings, D.Sc., Ph.D., C.Chem., F.R.S.C., is a graduate of the University of Birmingham, England with a B.Sc. in Chemistry (1953) and a Ph.D. (1956) for a thesis on ultrasonic degradation of polymers. After working for the Ministry of Aviation he became Senior Lecturer in Polymer Science at the University of New South Wales, Sydney, Australia (1961 to 1972). During this time he was Visiting Professor at Twente Technological University, Enschede, The Netherlands (1968-69), advising on their program in biomedical engineering. While in Australia, he became associated with Bernard Bloch, F.R.C.S., Orthopedic Surgeon, Sydney Hospital, and began a fruitful collaboration in the uses of plastics materials in surgery.

In 1972 he returned to England as Principal Lecturer in the Biomedical Engineering Unit of the North Staffordshire Polytechnic and the (now) North Staffordshire Health District with responsibility for research. With a particular interest in biomaterials research his own work has encompassed carbon fiber composites for surgical implants, adhesives, bioceramics, prosthesis performance in vivo, and electrical phenomena in bone. He is a member of British and International Standards Committees dealing with surgical implants and of other professional and scientific bodies, including Companion Fellow of the British Orthopaedic Association and Editor of the international Journal *Biomaterials*. He was elected President of the Biological Engineering Society in the U.K. (B.E.S.) in October, 1982. He was awarded a D.Sc. from the University of Birmingham in 1980 for a thesis in the field of biomedical applications of polymers. He has recently been appointed Acting Head of the department.

Paul Ducheyne, Ph.D. obtained the degree of metallurgical engineering from the Katholieke Universiteit Leuven, Belgium, in 1972. Subsequently he worked at the same university towards a Ph.D. on the thesis "Metallic Orthopaedic Implants with a Porous Coating" (1976). He stayed one year at the University of Florida as an International Postdoctoral N.I.H. Fellow and a CRB Honorary Fellow of the Belgian-American Educational Foundation. Thereafter he returned to the Katholieke Universiteit Leuven. There he was a lecturer and a research associate, affiliated with the National Foundation for Scientific Research of Belgium (NFWO). He recently joined the University of Pennsylvania, Philadelphia, as an Associate Professor of Biomedical Engineering and Orthopedic Surgery Research.

Dr. Ducheyne has published in major international journals on mechanical properties and design of prostheses, porous materials, bioglass, hydroxyapatite, and microstructural methods of analysis of biomedical materials. He is member of the editorial board of *Biomaterials, Journal of the Engineering Alumni of the University of Leuven, Journal Biomedical Materials Research,* and *Journal Biomechanics and Comtex System for Biomechanics and Bioengineering.*

He became active in various societies and institutions and has held or is holding the positions of Chairman-Founder of the "Biomedical Engineering and Health Care Group" of the Belgian Engineering Society, Secretary of the European Society for Biomaterials and member of the Board of Directors of Meditek (Belgian Institution to promote biomedical industrial activity).

CONTRIBUTORS

Oger F.. Arkens, Ph.D.
First Assistant
Department Metaalkunde
Katholieke Universiteit
Leuven, Belgium

P. De Meester, Ph.D.
President of Exact Sciences Group
Department Metaalkunde
Katholieke Universiteit
Leuven, Belgium

Robert H. Doremus, Ph.D.
Professor of Glass and Ceramics
Materials Engineering Department
Rensselaer Polytechnic Institute
Troy, New York

John H. Dumbleton, Ph.D.
Howmedica, Inc.
Rutherford, New Jersey

G. Heimke, Ph.D.
Friedrichsfeld GmbH
Mannheim, West Germany

Paul Higham, Ph.D.
Howmedica, Inc.
Rutherford, New Jersey

A. N. Hughes, Ph.D.
Principal Scientific Officer
A.W.R.E.
Reading, England

Ronald Kousbroek, M.Sc.
Delfzijl
The Netherlands

R. W. Lycett, Ph.D.
Ministry of Defense
Whitehall
London, England

Robert M. Pilliar, Ph.D.
Professor
Faculty of Dentistry
Department of Metallurgy and Materials
  Science
Toronto, Ontario
Canada

H. Richter, Ph.D.
Senior Scientist
Fraunhofer-Institut für Werkstoffmechanik
Freiburg, W. Germany

M. Semlitsch, Ph.D.
Department of Research and Development
Winterthur, Switzerland

U. Soltész, Ph.D.
Senior Scientist
Fraunhofer-Institut für Werkstoffmechanik
Freiburg, W. Germany

W. Van Raemdonck, M.Sc.
Department Metaalkunde
Katholieke Universiteit
Leuven, Belgium

TABLE OF CONTENTS

Volume I

Volume II

Chapter 1

MECHANICAL PROPERTIES OF SELECTED IMPLANT METALS USED FOR ARTIFICIAL HIP JOINTS

M. Semlitsch

TABLE OF CONTENTS

I. IMPLANT METALS IN CLINICAL USE

It is estimated that every day about 1000 artificial hip joints are implanted throughout the world in patients suffering from disabling hip joint disease in order to provide relief of pain and to restore or improve their joint function. This means replacing the femoral head and the acetabular socket with artificial prosthetic components so that motion then takes place on the articulating joint surfaces.

Iron-, titanium-, and cobalt-base alloys marketed under various trade names have been hitherto used in manufacturing artificial hip joints (Table 1). The chemical composition, structure, and mechanical properties of these alloys are described in national and international standard specifications.[20] These metallic materials are used for prosthetic components in combination with polymeric plastics (ultrahigh molecular weight polyethylene and polyoxymethylene) and aluminum oxide ceramic (Table 2).

II. CHEMICAL COMPOSITION OF IMPLANT METALS

On the basis of their chemical composition (Table 3), the clinically applied implant metals are iron-, titanium-, and cobalt-base alloys. The alloying elements contained in the metallic materials form homogeneous solid solutions of high corrosion resistance. The specific pure-element properties, such as the toxic and allergic reactions to cobalt, nickel, and vanadium,[15] become entirely insignificant in the alloy. This fact should be particularly stressed with regard to the well-known biocompatibility of these implant alloys with passive oxide films of good adherence at the implant surface.[11,30-32]

III. INFLUENCE OF PROCESSING METHODS ON THE MICROSTRUCTURE

The structure of an implant alloy is on one hand determined by its chemical composition and on the other hand by the manufacturing process used in the production of the alloy. Every alloy is primarily formed in a melting process, whereby this can be done in air, in an inert gas, or under vacuum. Because of the high requirements on the purity and mechanical strength of the alloy, preference is to be given to vacuum-melted materials.

The required implant shapes are produced by various methods. Casting is the most economical process, and with very complicated shapes it is generally also the only feasible method.[2]

Highly stressed components made in vacuum cast Co-Cr-Mo alloy may be subsequently also subjected to an isostatic hot-pressing process (HIP) in order to eliminate the pores and thus to increase its strength.[8,35] Another method of manufacturing implant components is the sintering process in combination with HIP of metal powders.[38] This process allows very homogeneous microstructures to be preserved at a high strength of the material. Because of the very high manufacturing costs, implants produced by powder metallurgy are very expensive.

Highly stressed implant components should be preferably manufactured by hot pressing and forging.[16] This extremely strength-increasing manufacturing process has been used for iron-, titanium-, and cobalt-base alloys for many years now.

Wrought Fe-Cr-Ni-Mo stainless steel — is the longest clinically used implant alloy. In the forged condition, this alloy shows an austenitic (face-centered cubic) crystalline structure. For hip prostheses, the low carbon quality is used either in the recrystallized condition (Figure 1A) or in the cold-worked state with about 50% deformation (Figure 1B). In order to increase the corrosion resistance of hip prostheses made of stainless steel, utmost attention must be given to the optimum shaping of the implant and the selection of a special quality.[1,29,41] A modified steel quality with an increased chromium and nitrogen content and a low niobium content has been recently introduced for Charnley hip prostheses.[26]

Table 1
IMPLANT METALS IN CLINICAL USE FOR ARTIFICIAL HIP JOINTS

ISO	Composition	Condition	Trade name	No.
5832-1	Fe-18Cr-14Ni-3Mo	Wrought	AISI-316 L	1
			AISI-316 LVM	
	Fe-21Cr-9Ni-4Mn-3Mo-Nb-N	Wrought	Ortron 90	2
5832-3	Ti-6A1-4V	Wrought	IMI-318A	3
			Protasul-64 WF	4
			Tioxium	5
			Tivaloy	6
			Tivanium	7
5832-4	Co-28Cr-6Mo	Cast	Alivium	8
			Endocast	9
			Orthochrome	10
			Orthochrome plus	11
			Protasul	12
			Protasul-2	13
			Vitallium cast	14
			Zimaloy	15
	Co-28Cr-6Mo	Wrought	Endocast hot worked	16
			Protasul-21 WF	17
			Vitallium FHS	18
	Co-28Cr-6Mo	P/M	Micro Grain	19
			Zimaloy	
5832-6	Co-35Ni-20Cr-10Mo	Wrought	Biophase	20
			MP-35N	21
			Protasul-10	22

Note: Trade names: American Society for Testing and Materials (1); Ceraver, France (5); DePuy U.S. (10,11); Fried. Krupp GmbH, BRD (9, 16); Howmedica Inc., U.S. (14, 18); Imperial Metal Industries, GB (3); OEC Orthopaedic Ltd. (6, 8); Richards Manufacturing Company, U.S. (20); SPS Technologies Inc., U.S. (21); Sulzer Bros. Ltd., Switzerland (4, 12, 13, 17, 22); Thackray Ltd., GB (2); Zimmer Inc. U.S. (7, 15, 19).

Table 2
COMBINATION OF IMPLANT MATERIALS IN USE FOR THE SOCKET, BALL, AND STEM OF ARTIFICIAL HIP JOINTS

Socket	Ball	Stem	Design
Co-Cr-Mo, cast	Co-Cr-Mo, cast	Co-Cr-Mo, cast	McKee-Farrar
Polymer	Fe-Cr-Ni-Mo, wrought	Fe-Cr-Ni-Mo, wrought	Charnley
Polymer	Co-Cr-Mo, cast	Co-Cr-Mo, cast	St. Georg Weller
		Co-Ni-Cr-Mo, wrought	Mueller Weber
		Co-Cr-Mo, wrought	Harris
		Ti-Al-V, wrought	Stanmore
Polymer	Ti-Al-V wrought	Ti-Al-V wrought	STH (Sarmiento)
Polymer	Al$_2$O$_3$, sintered	Co-Ni-Cr-Mo, wrought	Mueller Weber, Weber-Stuehmer
		Ti-Al-V, wrought	Zweymueller
		Fe-Cr-Ni-Mo, wrought	Shikita
		Co-Cr-Mo, cast	Lord
Polymer	Co-Cr-Mo, P/M	Co-Cr-Mo, P/M	TR-28 (Amstutz)
Al$_2$O$_3$, sintered	Al$_2$O$_3$, sintered	Ti-Al-V wrought	Boutin
		Co-Cr-Mo, cast	Mittelmeier

Table 3

CHEMICAL COMPOSITION OF IRON-, TITANIUM-, AND COBALT–BASE ALLOYS USED FOR ARTIFICIAL HIP JOINTS

Elements, Wt%	Fe-Cr-Ni-Mo ISO 5832/1	Ti-Al-V ISO 5832/3	Co-Cr-Mo ISO 5832/4	Co-Ni-Cr-Mo ISO 5832/6
Al, aluminum		5.50—6.75		
C, carbon	—0.03	0.08	—0.35	—0.025
Co, cobalt			Balance	Balance
Cr, chromium	16.0—19.0		26.5—30.0	19.0—21.0
Cu, copper	—0.50			
Fe, iron	Balance	—0.30	—1.0	—1.0
H, hydrogen		—0.015		
Mn, manganese	—2.0		—1.0	—0.15
Mo, molybdenum	2.0—3.5		4.5—7.0	9.0—10.5
N, nitrogen		—0.05		
Ni, nickel	10.0—16.0		—2.5	33.0—37.0
O, oxygen		—0.20		
P, phosphorus	—0.25			—0.015
S, sulfur	—0.015			—0.010
Si, silicon	—1.0		—1.0	—0.15
Ti, titanium		Balance		—1.0
V, vanadium		3.50—4.50		

Wrought Ti-Al-V alloy — has been used as an implant alloy since the 1950s.[12,38] This alloy used for the anchorage stems of total hip prostheses with a Al_2O_3 ceramic ball head has proven itself clinically in more than 10,000 cases since 1972.[4,5] The alloy is obtained with two phases (hexagonal α plus cubically body-centered β-phase) at forging temperatures in the β-range with a lamellar grain structure (Figure 2A) and in the α/β-range with a globular structure of Protasul-64WF® (Figure 2B). Lower deformation rates result in a nonequiaxed globular phase mixture (Figure 2C). Heat treatment at 700 to 750°C in vacuum does not alter these structures as seen in light microscopy.

Cast Co-Cr-Mo alloy — which was first employed for hip joint cups (cup arthroplasty) about 40 years ago,[28] is still the most commonly used alloy for total hip prostheses.[17] The carbon-containing alloy has an austenitic crystalline structure with interdendritically arranged block carbides of type $M_{23}C_6$ (M = Cr + Mo + Co) in the as-cast condition of Protasul® (Figure 3A) and in the heat-treated (below 1180°C) condition of Protasul-2® (Figure 3b). These block carbides, which are readily dissolved in the solution-annealed condition (heat treatment temperature approximately 1240°C), give rise to "Kirkendall holes" (Figure 3C).

Wrought Co-Cr-Mo or P/M alloy — is a further development of the well-known and clinically proven cast Co-Cr-Mo alloy.[3,6,10,14,21,22,24] As a result of hot forging and the powder metallurgy process, respectively, the alloy shows a higher density, a much smaller grain size, and a finer distribution of block carbides (Figure 4A and B). To attain maximum strength, this material can be hot-forged at optimum deformation rates without recrystallization of the austenitic structure. A forging process of this kind results in a longitudinal orientation of the austenitic grain structure of Protasul-21WF® (Figure 4C).

Wrought Co-Ni-Cr-Mo alloy — is the first high-strength implant alloy developed at the end of the 1960s[16,20] to be used for the manufacture of highly stressed anchorage stems of artificial joints of the hip (Mueller, Weber, and Weber — Stuehmer design), the knee (GSB design), the elbow (GSB design), and the hand (Meuli design). In the recrystallization annealed condition, the alloy shows a fine-grained, totally austenitic structure (Figure 5A). Hot forging above 650°C also produces a totally austenitic structure of Protasul-10® with elongated grains and a lattice dislocation configuration that is hardly movable after cooling

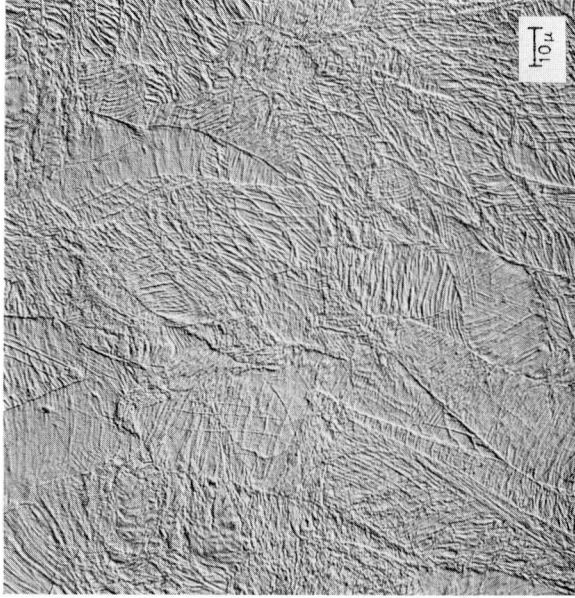

B

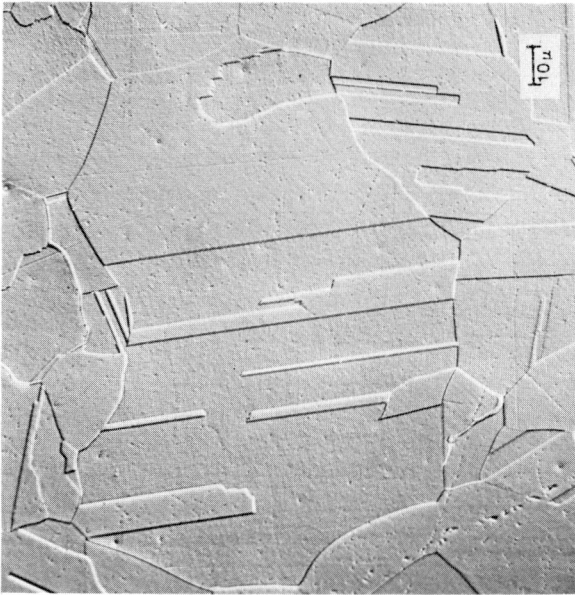

A

FIGURE 1. Microstructure of wrought Fe-Cr-Ni-Mo stainless steel AISI-316 L; (A) Austenitic grain structure in the annealed condition; (B) cold-deformed austenitic structure with elongated grains.

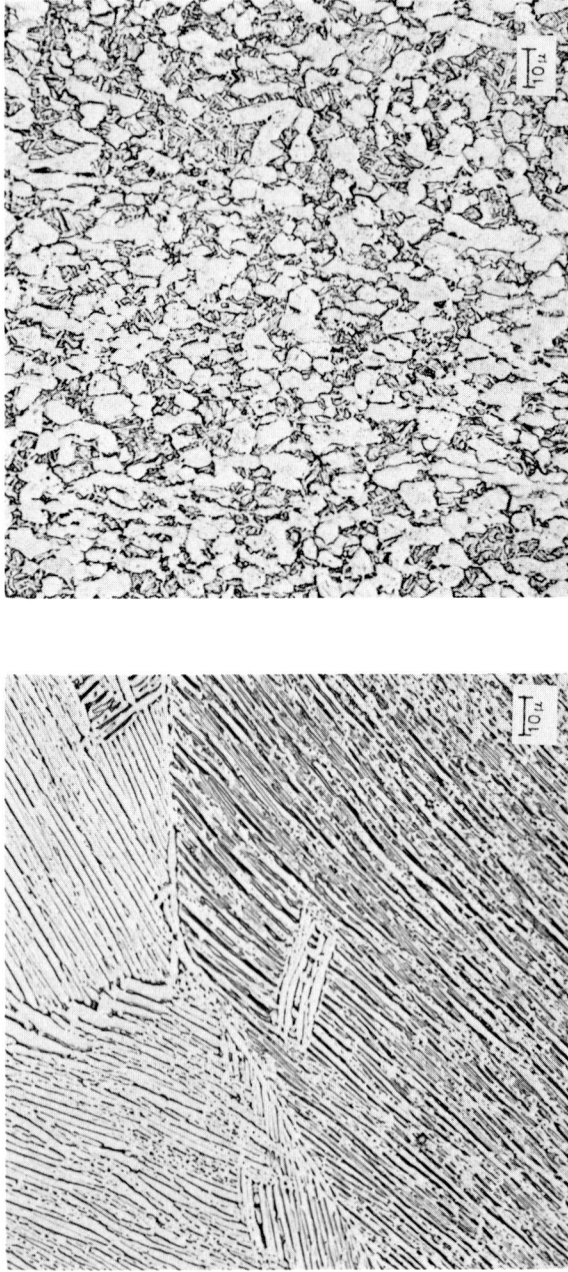

A

B

FIGURE 2.   Microstructure of wrought Ti-Al-V alloy. (A) Lamellar mixture of the α- and β-phase after hot forging in the β-range; (B) globular mixture of the hexagonal α-phase (light) and cubically body-centered β-phase in the hot-forged condition (Protasul-64WF® of globular quality); (C) nonequiaxed globular mixture of the α- and β-phase after hot forging in the α/β-range with a low deformation rate (Protasul-64WF® of nonequiaxed globular quality).

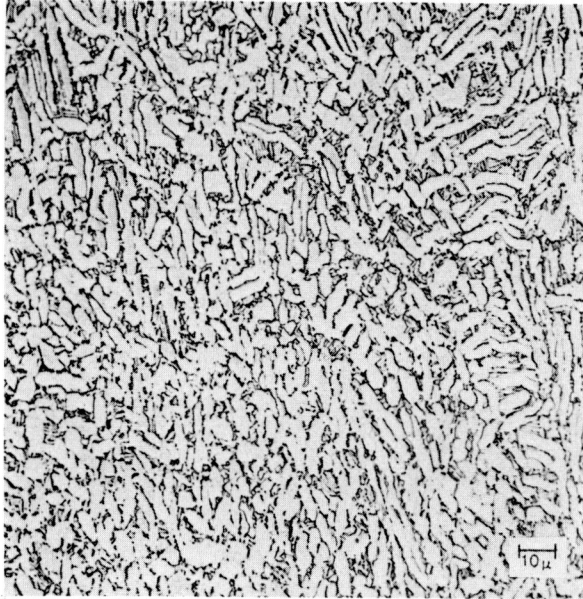

FIGURE 2C.

(Figure 5B). 50% Cold working enhances the lattice imperfection density of MP-35N® and the formation of an ϵ-martensite phase with hexagonal structure.[7,33] Subsequent heat treatment at 520 to 620°C results in precipitation of an intermetallic Co_3Mo phase (Figure 5C).

Cast Co-Cr-Mo/wrought Co-Ni-Cr-Mo double alloy — of TIG-welded composite construction[17] has been used since the early 1970s for the earlier-mentioned prosthesis types with wear-resistant cast components (Co-Cr-Mo Protasul® or Protasul-2®) and fatigue-resistant anchorage stems (Co-Ni-Cr-Mo Protasul-10®). Welding in an inert argon gas atmosphere results in an intimate combination of the two cobalt-base alloys (Figure 6A and B) of high corrosion resistance in the human body environment.

IV. MECHANICAL PROPERTIES UNDER STATIC AND DYNAMIC LOADS

Artificial hip joints implanted in the body are subjected to about 1 to 2 × 10^6 times a year by a multiple of the patient's body weight, hence, the requirement for implant materials of high mechanical strength under static and dynamic loading. These characteristic data are determined in static tensile tests and in dynamic rotating bending tests.[13] On the basis of clinical long-term experience with over 500,000 original standard Mueller and Weber total hip prostheses, stringent minimum requirements must be imposed on the strength of the metallic materials used with these total hip prosthesis types. The following strength values are therefore of vital importance in clinical application.

1. Yield strength as the measure of safety against permanent deformation of a body-loaded hip prosthesis stem (if an implanted femoral component stem is to be deformed only elastically and not plastically, a minimum yield strength of 450 N/mm² or higher is required.) (Figure 7)
2. Ultimate tensile strength as the measure of safety against the risk of fracture of a femoral component stem in case that it is subjected to a forced load (a minimum ultimate strength value of 800 N/mm² or higher is consequently required (Figure 8)

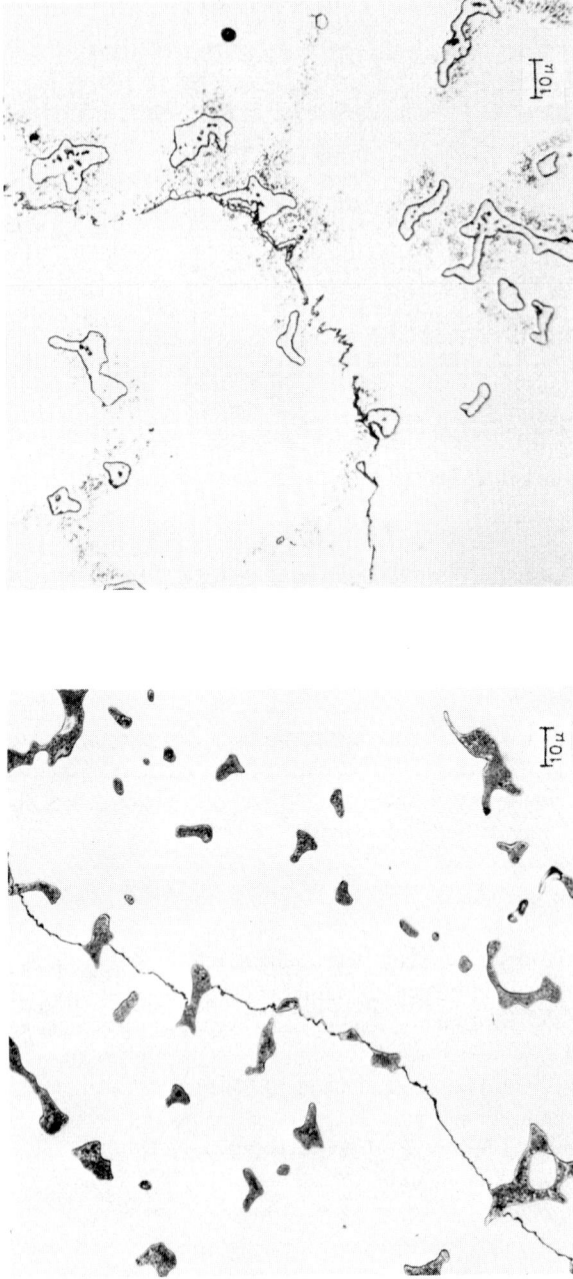

FIGURE 3. Microstructure of cast Co-Cr-Mo alloy. (A) Austenitic coarse-grained structure with eutectic block carbides (dark) in the as-cast condition (Protasul®); (B) the same microstructure as in Figure 3A after heat treatment at 1180°C (Protasul-2®); (C) austenitic grain structure with "Kirkendall holes" seen in the place of interdentritically arranged block carbides after solution annealing at 1240°C.

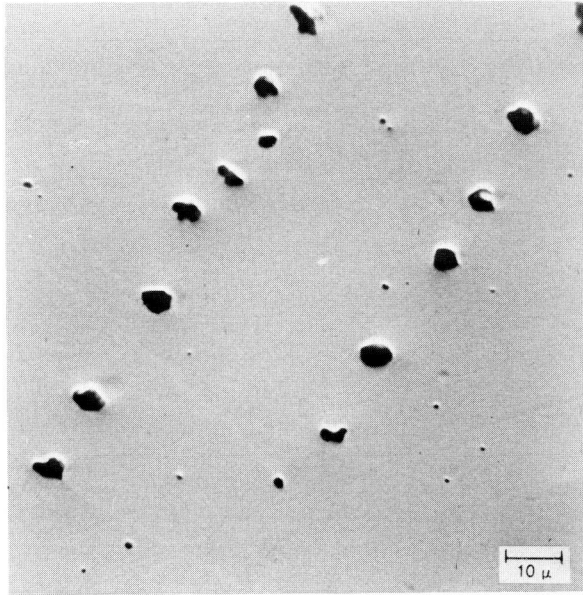

FIGURE 3C.

to protect the patient with an implanted hip prosthesis in the case of a fall. The ultimate tensile strength is, however, also decisive for the height of the fatigue strength (30 to 60% of the tensile strength) and is consequently also of great significance.

3. Elongation as the measure of the metallic material's toughness should be at least 8% or higher (Figure 9). If the orthopedic surgeon requires additional bending of the anchorage stem for optimum matching to the configuration of the bone, high elongation values are recommendable.

4. Fatigue strength as the measure of safety against fatigue fracture of a femoral component stem subjected to a millionfold load in bending and torsion in the body; this especially in the case of loosened anchorage stems. Minimum fatigue strength values of 400 N/mm^2 or higher are to be attained (Figure 10) to protect the anchorage stem of the femoral component against fracture. With fatigue strength values determined in air, allowance should be made for the fact that in the case of wrought Fe-Cr-Ni-Mo stainless steel a decrease in this value is to be reckoned with under the corrosive conditions in the human body environment.[27,36] This applies to cobalt- and titanium-base alloys to a minor extent only.

When comparing the fatigue strength values of implant alloys of different makes, the following points should be particularly taken into consideration:

1. Manufacturing process used in the production of the implant material, e.g., casting, cold working, hot forging or powder metallurgy process

2. Origin of the test specimens used for determining the mechanical strength values (the most conclusive values are obtained with specimens taken from the implant itself. Less conclusive are the data obtained with separately prepared test pieces, as is frequently the case with cast femoral prostheses.)

3. Surface condition (longitudinally polished or shot-peened) of flat or cylindrical test pieces used for determining the fatigue strength at 10 million load cycles with indication of the scatter values and confidence limits (It is generally known that shot peening

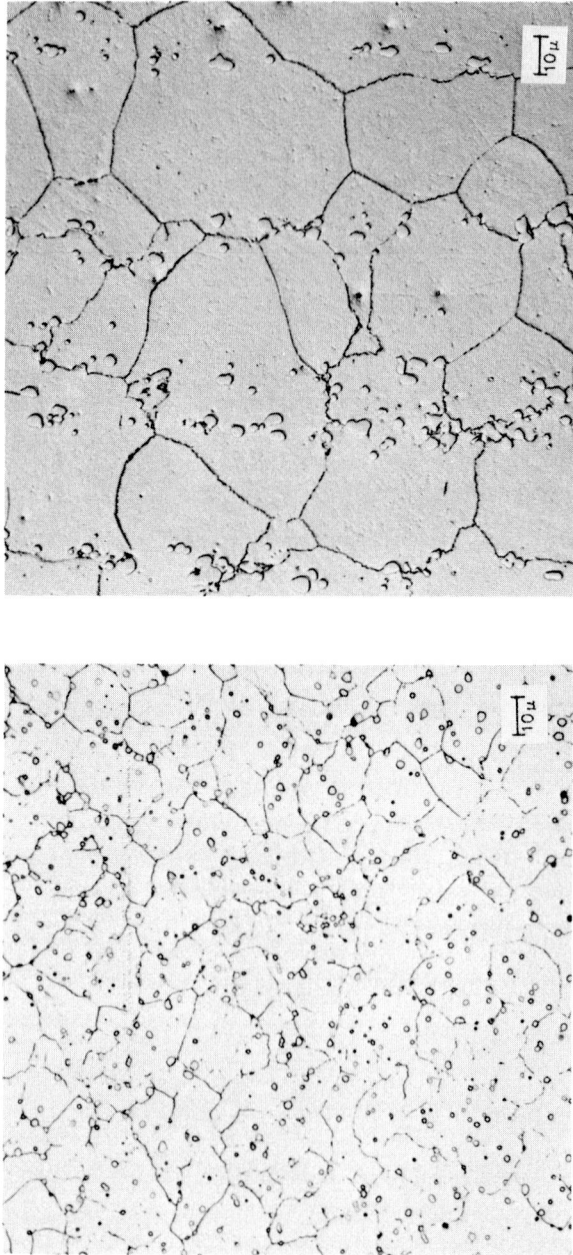

A

B

FIGURE 4.   Microstructure of high-strength Co-Cr-Mo alloy. (A,B) Austenitic fine-grained structure with a fine distribution of small block carbides in the annealed condition after the P/M process (Micro Grain Zimaloy®) and after hot forging (Protasul-21WF® of annealed quality); (C) hot-forged austenitic structure with elongated grains and finely distributed small block carbides (Protasul-21WF® of medium-hard quality).

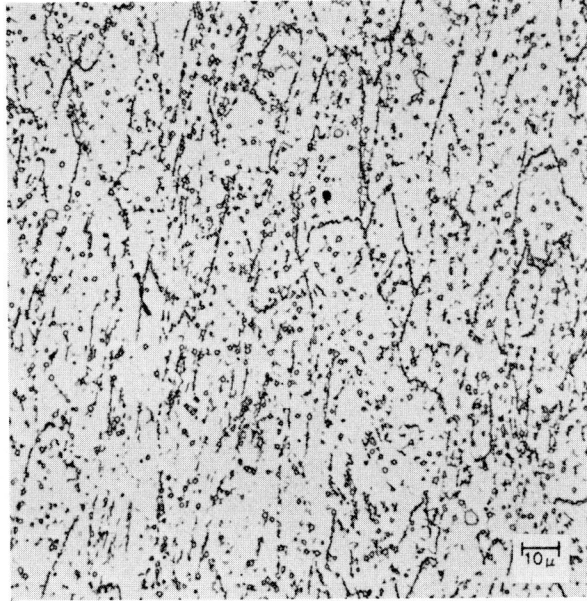

FIGURE 4C.

induces residual compressive stresses in the metallic material, whereby its fatigue strength can be increased by another 10 to 30%, as compared with mirror-finished test specimens) (Figure 11).

On the basis of the fatigue strength values obtained in rotating bending tests, the clinically applied metallic materials may be divided into two groups. The first group includes wrought Fe-Cr-Ni-Mo stainless steel and cast Co-Cr-Mo alloy, which — irrespective of the hitherto applied treatment conditions — never reaches a rotating bending strength of 450 N/mm^2. The second group with much higher values of up to 870 N/mm^2 includes wrought or P/M-produced Co-Cr-Mo alloy as well as Ti-Al-V and Co-Ni-Cr-Mo alloys in the hot-forged state.

V. CORROSION FATIGUE STRENGTH OF HIP REPLACEMENTS

When designing a hip prosthesis, the stresses expected in the human body should be known in order to be able to provide a safety factor against failure of the selected implant material. This applies in particular to loosened anchorage stems of the femoral component. For practice-oriented static and dynamic strength testing, the stem of the femoral component to be tested is inserted into a standard femur[34] in the neutral position and cemented in up to 50 mm below the collar (Figure 12). This allows severe loosening of the femoral component stem to be simulated. The distribution of stresses at the back of the stem is first determined in static bending tests by means of strain gauges (Figure 13). In this connection, the peak stresses at measuring point 3 are of major interest. For an original Mueller standard hip prosthesis made of hot-forged Co-Ni-Cr-Mo alloy Protasul-10® (yield strength = 1000 to 16000 N/mm^2) and subjected to a load of 5000 N (5 times the body weight of 100 kg), a bending stress of 1000 N/mm^2 was attained. Under this load, the stem shows a visible elastic deflection, but no plastic permanent deformation of the stem is registered. A Mueller hip prosthesis of the same design made of a material with a lower yield strength could not be loaded so highly. However, if the stem cross section to be loaded is increased, or if the

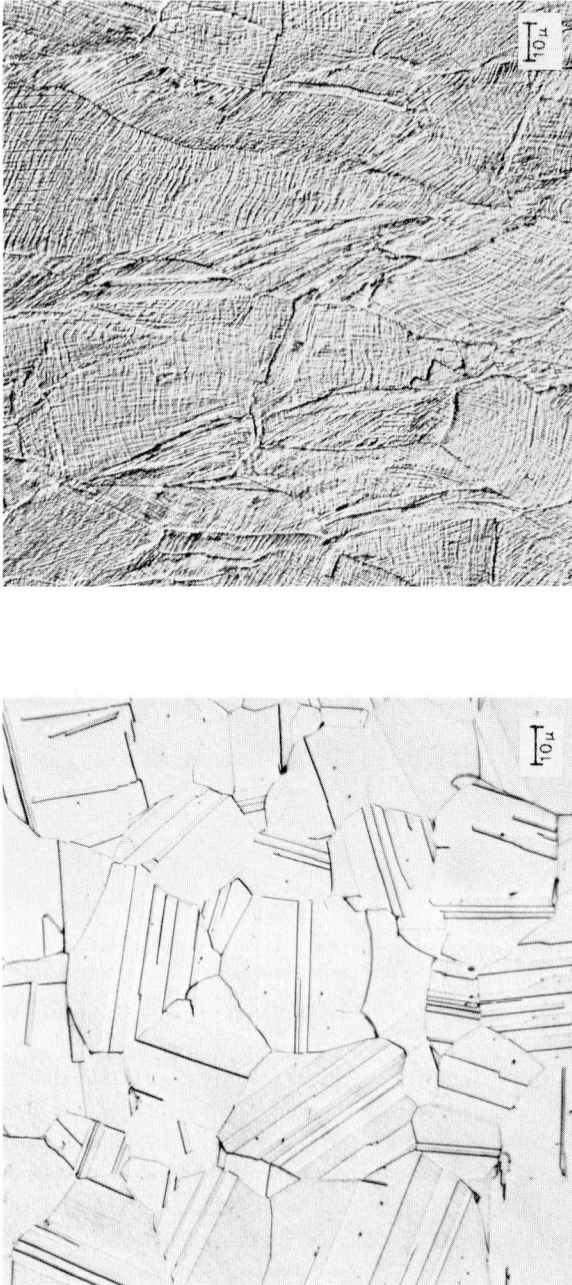

A

B

FIGURE 5.   Microstructure of wrought Co-Ni-Cr-Mo alloy. (A) Austenitic grain structure in the annealed condition after heat treatment at 1050°C (MP-35N®/Protasul-10® of annealed quality); (B) hot-forged austenitic structure with elongated grains and subgrains showing dislocations on a submicroscopic scale (Protasul-10® of hard quality); (C) 50% Cold-deformed austenitic structure with elongated grains, hexagonal ε-martensite, and precipitated intermetallic Co₃Mo phase after heat treatment at 520 to 610°C (MP-35N of extra-hard quality, Biophase®).

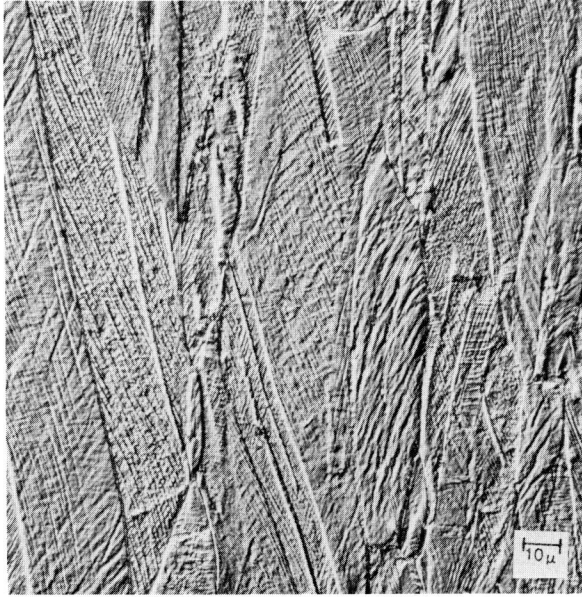

FIGURE 5C.

shape of the femoral prosthesis stem is generally changed (in the direction of lower peak stresses), the lower strength of the material (e.g., stainless steel or cast Co-Cr-Mo alloy) can be compensated through this.

As far as clinical application is concerned, it is ultimately the corrosion fatigue strength of the simulation-loosened femoral component stem (submerged in a corrosive air-fluxed Ringer's solution at 37°C) of up to 5 million cycles at a cyclic frequency of 5 to 10 Hz that is of major significance.[9,19] The latter test procedure, which has proven itself in research and production already since 1970 at Sulzer in Winterthur, may pave the way for international standardization. Determined is the maximum sustainable loadability of the simulation-loosened femoral component stem, whereby neither permanent deformation nor a fatigue crack or fracture of the prosthetic stem must occur.

The load limit of 5 million cycles has been settled upon on the basis of clinical experience with loosened femoral components which would — under similar conditions of loosening — never remain in the body for more than 1 year (1 to 2 million stress cycles). A femoral component stem that has been severely loosened for a period of several months to 1 year should certainly be able to withstand a severe overload without any failure. Consequently, this should contribute to preventing the removal of loosened femoral components under aggravated conditions involving risks to the patient and thus facilitate reoperation of loosened femoral prostheses that have not undergone any fatigue fracture.

The conclusiveness of this test method for the safety of a femoral prosthesis model against fracture is illustrated by the example of the Mueller hip prosthesis manufactured from two different implant alloys of different strength. Following the so-called staircase technique, as many as 10 hip prostheses are pulsated at varying loads of up to a maximum of 5 million cycles (some of them even up to 10 million cycles). To determine the corrosion fatigue strength, the ruptures obtained in this test and the pulsated prostheses without rupture are entered into a Woehler diagram (Figure 14).

At the same stem geometry of the Mueller hip prosthesis and under the same test conditions, this reveals that the safe limit of corrosion fatigue loading is of the order of 1400 *N* (corresponding twice the body weight of 70 kg) for annealed wrought Fe-Cr-Ni-Mo stainless

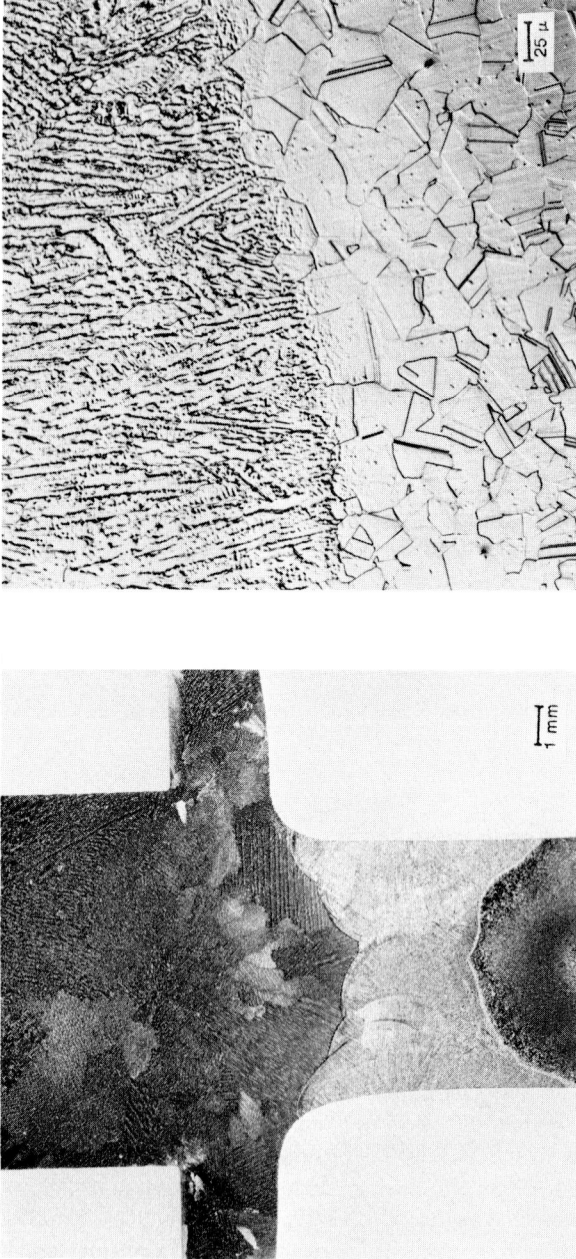

B

A

FIGURE 6. Microstructure of the TIG-welded combination of two cobalt-base alloys in the case of Weber total hip prostheses. (A) Articulating component (above) made of cast Co-Cr-Mo Protasul-2® alloy welded to a hot-forged Co-Ni-Cr-Mo Protasul-10® anchorage stem; (B) detail from Figure 6A showing the transition from the welding zone (above) to the recrystallized structure of the Protasul-10® alloy.

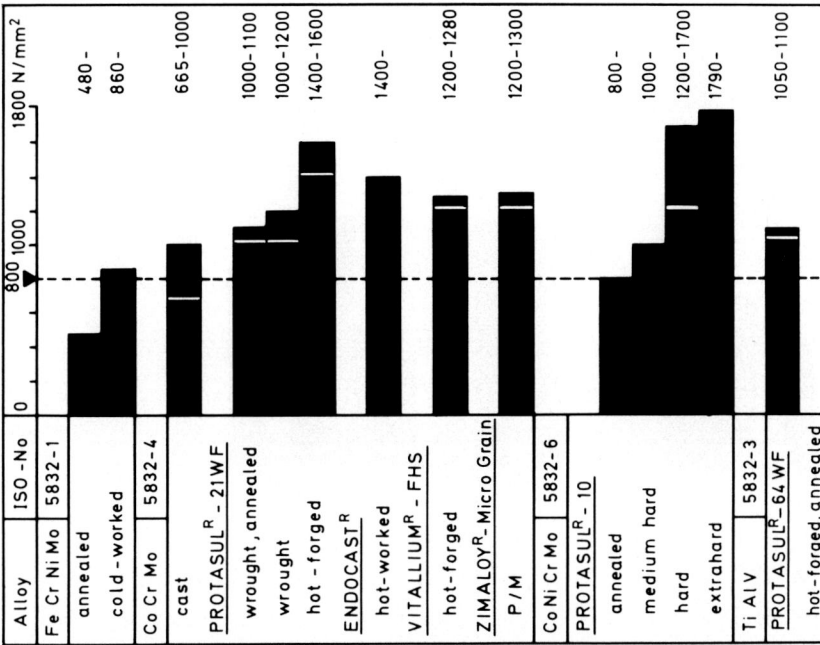

FIGURE 8. Ultimate tensile strength values of metallic implant materials used for artificial hip joints.

Alloy	ISO-No		
Fe Cr Ni Mo	5832-1		
annealed			480 –
cold-worked			860 –
Co Cr Mo	5832-4		
cast			665-1000
PROTASUL[R] – 21 WF			
wrought, annealed			1000-1100
wrought			1000-1200
hot-forged			1400-1600
ENDOCAST[R]			
hot-worked			1400-
VITALLIUM[R] – FHS			
hot-forged			1200-1280
ZIMALOY[R]-Micro Grain			
P/M			1200-1300
Co Ni Cr Mo	5832-6		
PROTASUL[R] – 10			
annealed			800 –
medium hard			1000-
hard			1200-1700
extrahard			1790-
Ti Al V	5832-3		
PROTASUL[R]-64 WF			
hot-forged, annealed			1050-1100

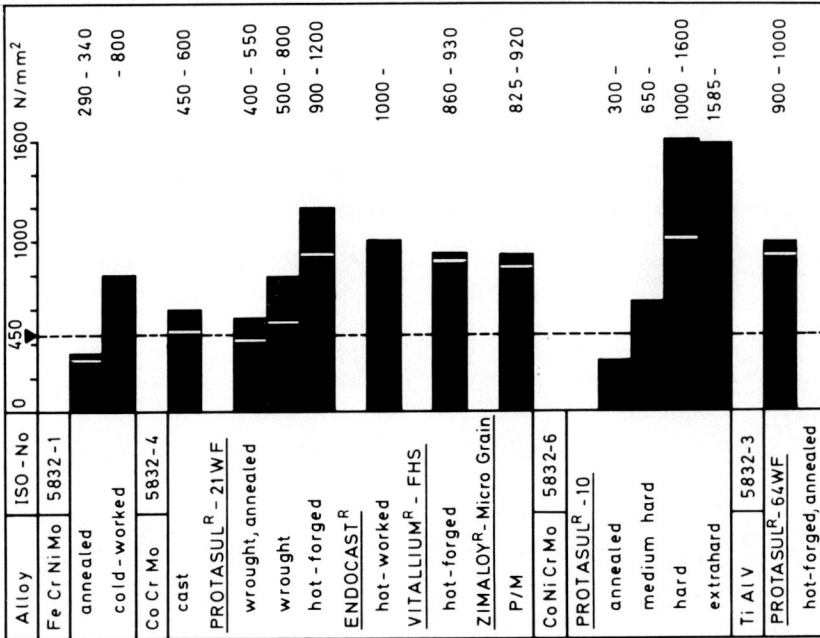

FIGURE 7. Yield strength values at 0.2% elongation of implant materials used for artificial hip joints.

Alloy	ISO-No		
Fe Cr Ni Mo	5832-1		
annealed			290 – 340
cold-worked			– 800
Co Cr Mo	5832-4		
cast			450 – 600
PROTASUL[R] – 21 WF			
wrought, annealed			400 – 550
wrought			500 – 800
hot-forged			900 – 1200
ENDOCAST[R]			
hot-worked			1000 –
VITALLIUM[R] – FHS			
hot-forged			860 – 930
ZIMALOY[R]-Micro Grain			
P/M			825 – 920
Co Ni Cr Mo	5832-6		
PROTASUL[R] – 10			
annealed			300 –
medium hard			650 –
hard			1000 – 1600
extrahard			1585 –
Ti Al V	5832-3		
PROTASUL[R]-64 WF			
hot-forged, annealed			900 – 1000

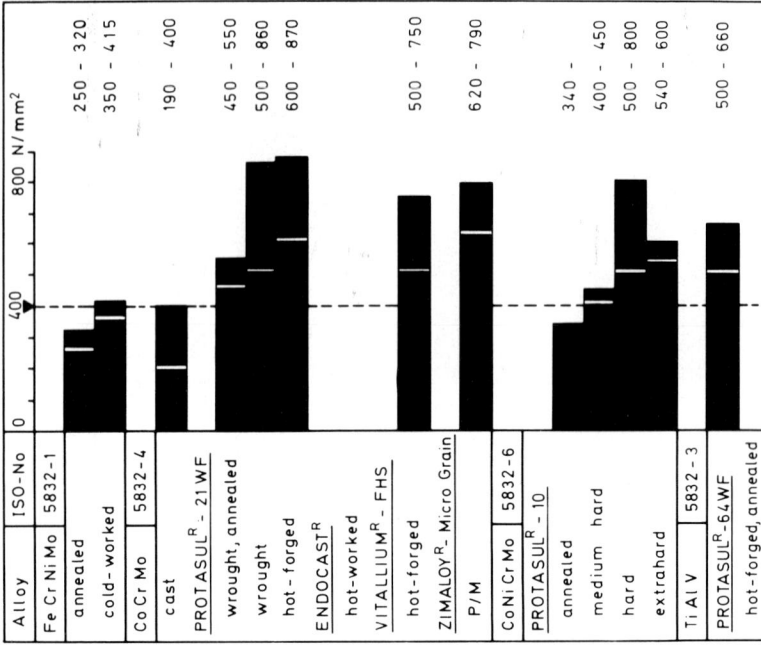

FIGURE 10.   Fatigue strength values of metallic implant materials used for artificial hip joints.

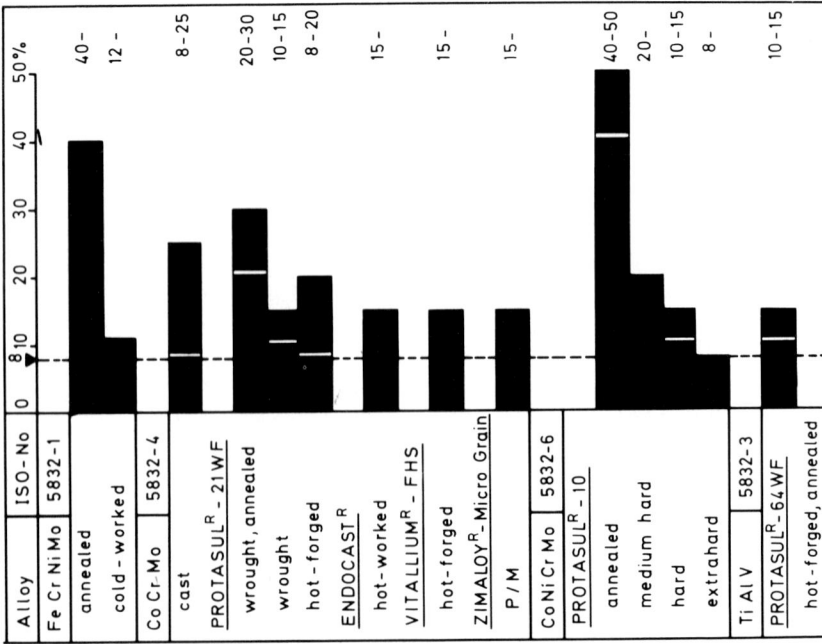

FIGURE 9.   Elongation values of metallic implant materials used for artificial hip joints.

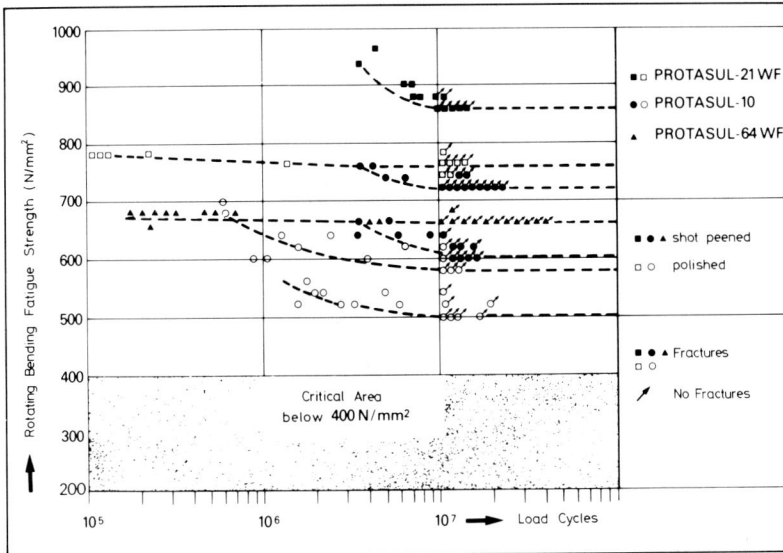

FIGURE 11.   Woehler curves of three implant alloys, Protasul-10®, Protasul-21WF®, and Protasul-64WF®, in the polished and shot-peened surface condition.

steel AISI-316 L and of the order of 4000 *N* (corresponding 5 times the body weight of 80 kg) for the hot-forged Co-Ni-Cr-Mo alloy Protasul-10® of hard quality.

In the light of many years of clinical experience gained with 500,000 original Mueller and Weber hip prostheses made of cast Co-Cr-Mo alloy and hot-forged Co-Ni-Cr-Mo alloy Protasul-10®, a minimum corrosion fatigue loading of 2800 *N* (3.5 times the body weight of 80 kg) should be attained today to be able to provide a guarantee of safety against fracture for loosened femoral component stems. The critical range of pulsating corrosion fatigue strength lies at 2000 *N* (under the specified test conditions). As a matter of fact, not a single case of failure (plastic deformation, fatigue crack, fracture, or corrosive attack) has been reported in connection with more than 400,000 series-produced original Mueller and Weber total hip prostheses with stems made of hot-forged Co-Ni-Cr-Mo alloy Protasul-10® of medium hard and hard quality, supplied since 1972.[13,20,37,40]

In static bending and pulsating fatigue strength tests, prosthesis stems made of hot-forged Ti-Al-V alloy show a much more marked deflection than hip prostheses of the same cross section made of cobalt-base alloys. This is explained by the approximately half as low modulus of elasticity of Ti-Al-V alloy (about 110,000 *N*/mm²) as compared with that of cobalt-base alloys (about 230,000 *N*/mm²). This is to be duly considered when designing implants in titanium alloy, because a titanium implant of the same cross section is only half as stiff as an implant made in a cobalt-base alloy. Whether this has a positive or negative effect on the bone structure as regards biomechanics has to be cleared up in each case. With femoral prostheses made in Ti-Al-V alloy, it is in particular the manufacturing technology that has a great influence on the pulsating corrosion fatigue strength of the femoral component stem.

VI. NEW DEVELOPMENTS — PROSPECTS FOR THE FUTURE

Modern materials engineering is now developing metallic, highly corrosion-resistant, fiber-reinforced superalloys, e.g., for thermally high-stressed gas turbines, and metallic materials are being partly replaced by fiber-reinforced polymeric plastics or carbon fiber-reinforced

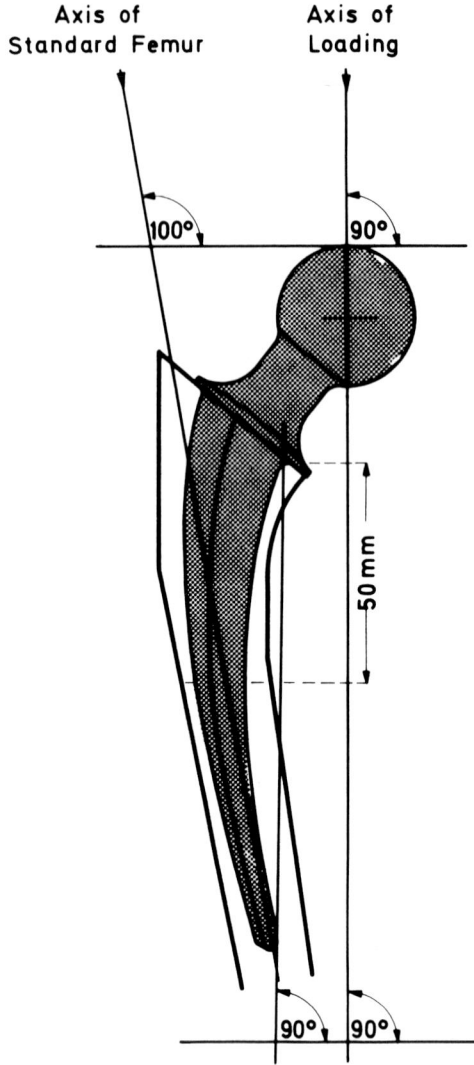

FIGURE 12.   Position of a standard Mueller hip prosthesis (model 10.32.20) in the standard femur.[34] Unce-
mented stem length, 50 mm; simulation of a loosened
stem condition.

carbon. The 1980s and 1990s will have to show to what extent materials derived from these
are suitable for implants, and in particular for hip endoprostheses. In the past 10 years,
combinations of metals, plastics, and ceramics, as well as of some coating materials, have
gained acceptance in the construction of artificial joints, and they can be hardly dispensed
with today. Experience has shown that it takes up to 10 years to develop a new implant
material and to clinically evaluate the first prototype series. This explains why relatively
few alloys are currently available for implants.

The future of implant alloys should be further considered also in the light of the scarcity
of raw materials as well as the steadily rising price trend of strategically important elements
such as cobalt, chromium, nickel, molybdenum, and titanium. Since the safety of long-term

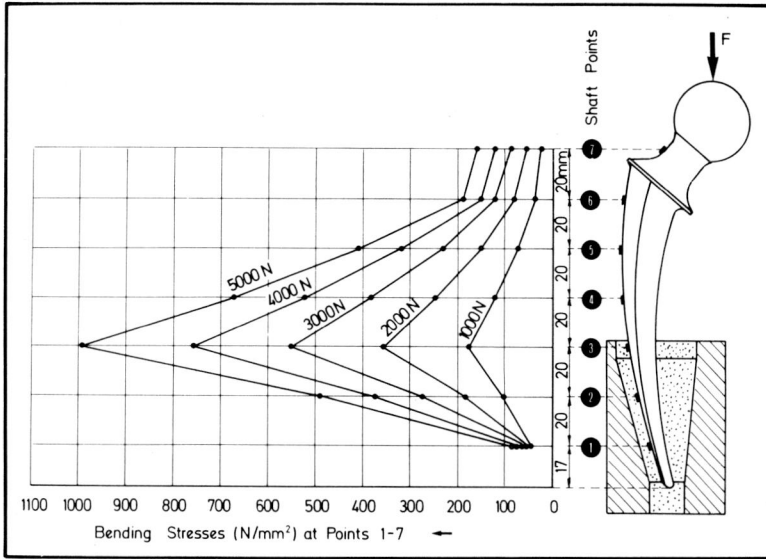

FIGURE 13.   Distribution of bending stresses at measuring points 1 to 7 for a standard Mueller hip prosthesis (Model 10.32.20) having a surface-structured stem made of wrought Co-Ni-Cr-Mo Protasul-10® alloy (hard-quality) under static loads of 1000 to 5000 N.

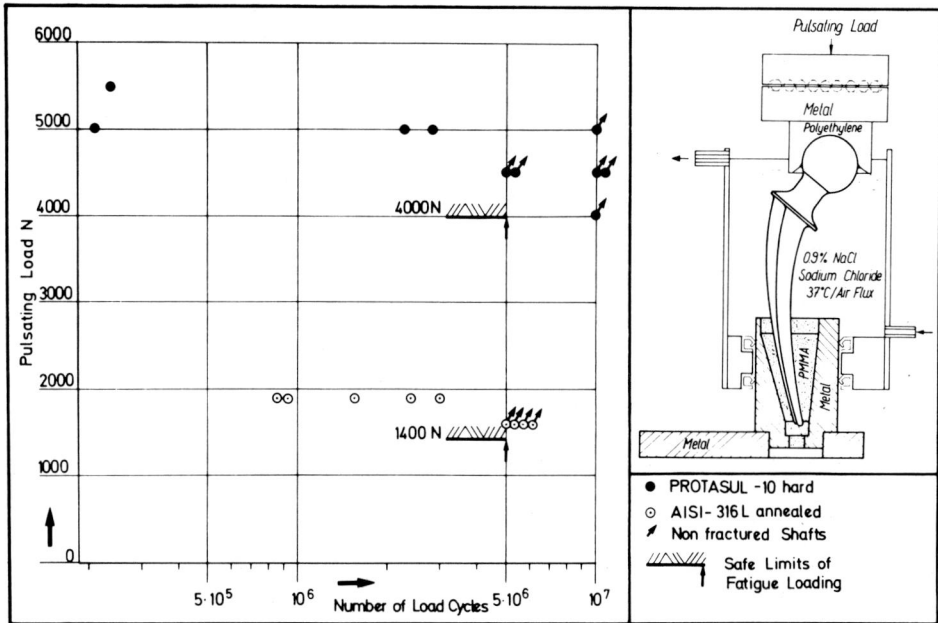

FIGURE 14.   Safe limits of functional fatigue loading F for standard Mueller hip prostheses (Model 10.32.20), made of hot-forged Co-Ni-Cr-Mo Protasul-10® alloy, and for Mueller hip prostheses of the same configuration, made of wrought Fe-Cr-Ni-Mo stainless steel AISI-316 L.

implant is of major concern both to the patient and the surgeon, only the best implant materials should be chosen for designing and manufacturing hip replacements. Price considerations at the expense of quality in respect of the implant itself should merely play a secondary role when considering the total cost for hip joint surgery, for this would mean economizing in the wrong place.

REFERENCES

1. **Bäumel, A.**, Spaltkorrosion an Implantaten, in *Implantatbrüche*, Deutscher Verband für Materialprüfung, Berlin, 1980, 73.
2. **Batt, I.**, *Investment Casting Handbook*, Investment Casting Institute, Chicago, 1968.
3. **Bensmann, G. and Müller, M.**, Properties and applications of workable Ni-free Co-base alloy, Transactions of World Biomaterials Congress, Baden, Austria, 1980, 2.29.
4. **Boutin, P. and Blanquaert, D.**, New materials used in total hip replacement, *Cah. Enseignement SOFCOT*, 10, 27, 1979.
5. **Christel, P.**, private communication, Goeteborg, Sweden, August 1981.
6. **Devine, T. M. and Wulff, I.**, Cast versus wrought cobalt-chromium surgical implant alloy, *J. Biomed. Mater. Res.*, 9, 151, 1975.
7. **Graham, A. H.**, Strengthening of multiphase alloys during aging at elevated temperatures, *Trans. Am. Soc. Met.*, 62, 930, 1969.
8. **Hodge, F. G. and Lee, T. S.**, Effects of processing on performance of cast prosthesis alloy, *Corrosion-Nace*, 31, 111, 1975.
9. **ISO**, TC-150, SC-1, WG-3, Corrosion Fatigue Testing of Hip Joint Endoprostheses, SNV-Draft Proposal, No. Ch-1/18, International Standardization Organization, Geneva, 1975.
10. **Kesh, A. K. and Kummer, F. J.**, New manufacturing and processing techniques for the fabrication of cobalt-chromium surgical implants, Trans. 24th Annu. ORS, Dallas, February 1978, 311.
11. **Kuehne, D.**, Die Gewebsverträglichkeit verschiedener metallischer Implantat-Werkstoffe in Pulverform, Dissertation, Orthopädische Univ. Klinik, Frankfurt a.M., Sulzer Reprint No. 28.75.00 Dgc 10, 1975, 1.
12. **Laing, P. G.**, private communication, May 1981.
13. **Lorenz, M., Semlitsch, M., Panic, B., Weber, H., and Willert, H. G.**, Fatigue strength of cobalt-base alloys with high corrosion resistance for artificial hip joints, *Eng. Med.*, 7, 241, 1978.
14. **Luckey, H. A. and Barnard, L. I.**, Improved properties of CoCrMo alloy by hot isostatic pressing of powder, Transactions 24th Annu. ORS, Dallas, February 1978, 296.
15. **Rae, T.**, The toxicity of metals used in orthopaedic prostheses, *J. Bone Jt. Syrg.*, 63B(3), 435, 1981.
16. **Semlitsch, M.**, Implant metals for plates, screws and artificial joints in bone surgery, *Sulzer Tech. Rev.*, 54, 245, 1972.
17. **Semlitsch, M.**, Technical progress in artificial hip joints, *Sulzer Tech. Rev.*, 4, 235, 1974; *Total Hip Prosthesis*, Hans Huber, Bern, 1976, 256.
18. **Semlitsch, M. and Willert, H. G.**, Cast and wrought cobalt base alloys as implant materials, *Med. Orthop. Tech.*, 96, 86, 1976.
19. **Semlitsch, M.**, Metallurgical and clinical experience with cast and forged cobalt-chromium base implant metals of compound construction for artificial joint endoprostheses, in *Reconstruction Surgery and Traumatology*, Chapchal, C., Eds., Karger, Basel, 1976, 82.
20. **Semlitsch, M.**, Properties of wrought CoNiCrMo alloy Protasul-10, a highly corrosion and fatigue resistant implant material for joint endoprostheses, *Eng. Med.*, 9(4), 201, 1980.
21. **Semlitsch, M. and Panic, B.**, Müssen Implantatmetalle heute noch brechen? Interdisziplinäres Symposium über Osteosynthese und Endoprothese, Winterthur, Switzerland, October 1980.
22. **Semlitsch, M. and Willert, H. G.**, Properties of implant alloys for artificial hip joints, *Med. Biol. Eng. Comput.*, 18, 511, 1980.
23. **Semlitsch, M. and Willert, H. G.**, Korrosions- und Festigkeitseigenschaften metallischer ISO-5832 Implantatwerkstoffe auf Eisen-,Kobalt- und Titanbasis für künstliche Hüftgelenke, *Symposium Biomaterials*, Gentner Verlag, Stuttgart, 1981, 54.
24. **Semlitsch, M. and Willert, H. G.**, Biomaterialien für Implantate in der orthopädischen Chirurgie, *Medizintechnik*, 101(3), 66, 1981.

25. **Semlitsch, M., Panic, B., Weber, H., and Schoen, R.,** Comparison of the fatigue strength of femoral prosthesis stems made of forged TiAlV and cobalt-base alloys, ASTM Symposium, Phoenix, May 1981.
26. **Smethurst, E.,** A new stainless steel alloy for surgical implants compared to 316 S 12, *Biomaterials,* 2,(2), 1981.
27. **Smith, C. J. E. and Hughes, A. N.,** The Influence of Frequency and Cold Work on the Fatigue Strength of 316 Stainless Steel in Air and 0.17 *M* Saline, Report number AWRE/44/83/189, UK Atomic Weapons Research Establishment, Aldermaston, England, 1977.
28. **Smith-Petersen, M. N.,** Evolution of mould arthroplasty of the hip joint, *J. Bone Jt. Surg.,* 30B(1), 59, 1948.
29. **Steinemann, S. G.,** Corrosion of surgical implants — in vivo and in vitro tests, in *Evaluation of Biomaterials,* Winter, G. D., Leray, J. L., and de Groot, K., Eds., John Wiley & Sons, New York, 1980.
30. **Suery, P.,** Untersuchungen zum Korrosionsverhalten gegossener und geschmiedeter Implantat-Werkstoffe, *Werkst. Korros.,* 26, 278, 1975.
31. **Suery, P.,** Korrosionseigenschaften gegossener und geschmiedeter Werkstoffe für künstliche Gelenke, *Medita,* 6, 19, 1976.
32. **Suery, P.,** Corrosion behavior of cast and forged cobalt-based alloys for double-alloy joint endoprostheses, *J. Biomed. Mater. Res.,* 12, 723, 1978.
33. **Treharne, R. W.,** The development of a new ultra high strength femoral implant material (Biophase), Transactions of World Biomaterials Congress, Baden, Austria, 1980, 2.28.
34. **Ungetheum, M.,** *Technologische und biomechanische Aspekte der Hüft-und Kniealloarthroplastik,* Hans Huber, Verlag, Bern, 1978, 1.
35. **Wasielewski, G. F. and Lindblad, N. R.,** Elimination of Casting Defects Using HIP, Proc. 2nd Int. Conf. on Superalloys-Processing, Seven Springs, Pa., September 1972.
36. **Wheeler, K. R. and James, L. A.,** Fatigue behavior of type 316 stainless steel under simulated body conditions, *J. Biomed. Mater. Res.,* 5, 267, 1971.
37. **Willert, H. G. and Semlitsch, M.,** Biomaterialien und orthopädische Implantate, in *Orthopädie in Praxis and Klinik II,* Georg Thieme Verlag, Stuttgart, 1981.
38. **Williams, D. F.,** Titanium as a metal for implantation, *J. Med. Eng. Technol.,* July 1977, 195.
39. **Williams, D. L.,** Hot isostatically pressed alloy APK 1, a nickel-base superalloy, *Powder Metall.,* 2, 84, 1977.
40. **Zichner, L. and Willert, H. G.,** Clinical experience with Mueller total hip endoprostheses with stems made of wrought CoNiCrMoTi alloy Protasul-10, Paper 93, Trans. 3rd Annu. Meeting, Society for Biomaterials, New Orleans, 1977, 98.
41. **Zitter, H. and Oberndorfer, M.,** Ursachen von Schadensfällen metallischer Implantate, in *Implantatbrüche, Deutscher Verband für Materialprüfung,* Berlin, 1980, 119.

Chapter 2

MECHANICAL BEHAVIOR OF SELECTED CERAMICS

U. Soltész and H. Richter

TABLE OF CONTENTS

I. INTRODUCTION

Although ceramics are the oldest artificial man-made materials, they were taken into consideration for biomedical applications as the last of the various material groups. Except for the use of plaster of Paris for extracorporal casts and for some limited attempts to fill defects in bone,[1] which seemed not to be successful in all cases,[2] the systematic and detailed evaluation of ceramics as possible materials for artificial organs of bone and joint replacement devices did not start before the 1960s. At that time metals and plastics had been in use for several decades. There were two main reasons for disregarding ceramics as implant materials. Technical ceramics are not very pure so that a good biocompatibility was uncertain, and — more decisively — they are all restricted in their strength compared with the other materials used. However in the late 1950s, with respect to new applications, different ceramics were improved in purity and strength. Additionally new favorable material properties of some ceramics like high wear resistance and low friction were recognized. Furthermore in the early 1960s surprising features in biological environments were discovered, from an almost complete inertness for some materials up to an active behavior which leads to bonding with the surrounding tissue or to degradation and conversion into bone.

Because of the specific requirements for "biomaterials", only some small sections out of the wide pallet of ceramic materials can be taken into consideration for implantations, of course, and only some of them have been selected for different applications and have been evaluated or are under investigation up to now. Following the classification suggested by Heimke et al.[2,3] these "bioceramics" can be divided into four groups characterized by their behavior in biological environments.

The first group comprises carbon in different modifications due to processing.[4] These materials will be treated in the next chapter, so that in this part only some data will be given to show the range for comparison with the other ceramics. The principal advantage of carbon materials is that they seem to behave completely inert or biocompatible. This is true not only for bone and soft tissue, but also for blood. Therefore, they can be used also in the cardiovascular system.

In contrast to this, all other ceramics investigated can be called biocompatible only in a restricted sense. They are only "bone and soft tissue compatible" or even only "bone tissue compatible". However, on the other hand they exhibit some other advantages.

There is a second group of "bioinert" materials which includes various oxides, nitrides, and carbides of base metals used in dense as well as in porous form. Some of them approach the definition of inertness rather perfectly, e.g., Al_2O_3-ceramics;[5-10] the others like MgO, stabilized ZrO_2, Si_3N_4, SiC, and Si-Al-ON seem to show a similar behavior.[2,3,5,11-15,115] This group comprises simultaneously the materials with the highest densities, Young's moduli, hardnesses, and strengths of all ceramics investigated for biomedical purposes. Most of them also show very good wear and gliding behavior.[16-22]

Except for Al_2O_3 all these materials are still in the phase of biocompatibility investigations at present; real applications for implant devices are not known. The alumina, however, is one of the most extensively tested materials and has been employed for different bone replacements for a few years. Because of its relatively high strength and good wear resistance, it has been chosen mainly for load bearing devices, e.g., as for hip joint components,[23-27] but because of its inertness it is used also for some other applications for which these favorable mechanical properties are not so important, as for dental implants,[28,29] in maxillofacial surgery,[30] and for keratoprostheses and ossicular chain replacements.[2]

These inert ceramics will be considered in this chapter in more detail. Especially in their dense modification they seem to provide further application possibilities.

The third group can be described as "bioactive" ceramics. It consists of different glasses or glass ceramics[31-35] or of rather dense calcium phosphate-based materials with a composition and structure similar to the inorganic components of bone.[36] These materials are characterized by a certain solubility which provokes the surrounding bone or tissue to form a direct bonding to the implant. This bonding is able to transfer also shear and tensile stresses along the interface,[34-39] which could be an advantage for anchoring the implants and reducing stress peaks in the bone.[40,41] The main restriction for these materials lies in their low strengths so that they can be used as bulk materials only for low loaded devices.[38,42] On the other hand, different attempts have been made in orthopedic and dental surgery to use them for coatings, as continuous layers,[43-45] as well as by embedding particles in enamels.[46,47] At least the latter seem to be successful, whereas the continuous glass layers cause some doubts because dissolution or partial loosening of the coating and inflammation were observed.[2,45] These materials will be discussed in further chapters especially under the aspect of their surface activity; therefore, only the data available from the literature for the bulk material will be given here for comparison with the inert ceramics.

The same shall apply to the fourth group, the so-called "biodegradable" ceramics. They are mainly based on calcium phosphates in different modifications,[2,36] besides the more historically interesting plaster of Paris ($CaSO_4$). They differ from the bioactive materials by a higher grade of solubility which leads to a gradual degradation and resorption by the surrounding tissue, stimulates the bone to grow on the material and through its pores, and is believed in some cases to generate a total transformation of the material into living bone in this way.[2,36,48-54] This means these ceramics could offer new possibilities in reconstructive surgery to fill or bridge bone defects. Therefore, they have found great interest in the last years and some remarkable progress has been made in improving their mechanical properties which depend strongly on the chemical structure of the various phosphates, their composition, on the porosity, and different processing conditions. Nevertheless, they seem to be limited in their applicability because of their strengths which are low for the bulk materials, and all the worse for the preferentially used porous forms. Thus, they can be applied only to low-loaded devices, e.g., as in maxillofacial surgery or for dental implants.[49,55] Some attempts have been made also to use these ceramics as coatings for higher strength materials,[56] but this seems not to be the right way for long-term implants since the coating will degrade gradually so that the substrate material will be exposed more and more to the surrounding tissue and the anchoring function will be lost.[36]

For the selection of a material for some particular purpose, besides its compatibility, the mechanical properties provide decisive criteria. Young's moduli and Poisson's ratio describe the deformability of the material and therefore influence the stiffness of a structure made of it. Their knowledge is important to evaluate the interaction between implant and surrounding tissue, i.e., the stress distributions which arise in the bone by loading the whole system bone-implant, and the stresses in the interface between both, in order to prevent resorption of the tissue or destruction of the interface connection by overstressing. The selection criterion can certainly not be a simple adjustment of these mechanical properties to those of the removed living material. This will only be true if the artificial replacement is equal in size and shape to the removed part and, moreover, is connected to its environment in exactly the same way with respect to the loading transfer. This means it could be correct, e.g., for ingrowing bridges. If, however, boundary conditions and shape are different, the whole mechanical situation will be changed and will have to be evaluated. In some cases on the contrary, it could be an advantage to use a very stiff structure and therefore a high modulus material in order to reduce stress peaks in the bone.[57,58]

The different ceramics available offer a wide variety of moduli for various purposes. Most of them behave in a nearly linear elastic manner over the whole possible loading range; the high-density ones accomplish this in an ideal manner. Therefore, plastic or viscoelastic influences and other nonlinearities can be neglected, or in other words, flow, creep, or relaxation need not be regarded.

An even more important criterion for applicability is the strength behavior because failure has to be prevented in any case. This demand restricts many of the ceramics in use. The reliability has not only to be guaranteed for some maximum possible load, but for the whole time of use. Therefore, the time dependence of the strength for the different possible loading conditions and the influence of the biological environment must be known on a long-term basis, or in other words, static and dynamic fatigue and corrosion behavior have to be considered. Furthermore, because of the high risk in human applications, a maximum possible safety, or a minimum probability for failure has to be ensured.

Most data available are stress-related mean values obtained under quasi-static loading conditions in different test arrangements. Such data, however, can give only an idea of the range in which the strength of the material concerned lies, since the measured values depend in general on the size, shape, and processing of the specimens, the test configuration, and the loading rate and show mostly a considerable scatter and do not admit any reliable conclusion on the long-term and corrosion behavior. For the special requirements on safety in biomedical applications, a more detailed description has to be postulated. Besides some information on the fatigue behavior under physiological conditions, the test conditions should be known in order to be able to correct the data for the assessment of the real structure, and at least, a statistical error should be declared which allows estimation of the probability for failure. A better description in the latter case would be that obtained by special statistical distributions, e.g., by Weibull statistics, which is applied more and more in the field of ceramics and has been proven to be useful for predicting the failure probability at certain loading levels.

The best characterization of strength, however, which is possible today, seems to be given by the concept of fracture mechanics. In this concept the strength is defined not only by stress-related values, but additionally by the size and shape of flaws, which exist unavoidably in each material or each component due to processing and influence the strength severely. Ceramic materials especially are suitable for this concept, since they behave in a linear elastic manner and brittle and fracture mechanics can be applied in its relatively simple, linear elastic form. This material behavior includes mostly a defined crack propagation behavior under constant environmental conditions, so that moreover predictions about the long-term properties become possible. One problem in applying this concept to ceramics could be the determination of the initial flaws. As far as is known, there does not exist any suitable technique to recognize and measure the small flaws in the critical range which are typical of ceramic materials. However, again by utilizing the fracture mechanics concept, the maximum possible flaws in a structure can be evaluated by a so-called "proof-test" in a proper loading arrangement. From these maximum possible flaws, minimum loading levels can be estimated which will be sustained without failure or minimum lifetimes can be calculated which will be achieved under given loading conditions. Thus, a theoretically absolute safety against failure can be guaranteed for known application cases which would fulfill the extreme safety requirements for biomedical use. This concept has been proved to be advantageous in the last years and has been applied in the meantime to characterize some dense ceramics used in endoprosthetic devices. It will be discussed in more detail in the following.

Besides the most important mechanical properties, strength and Young's modulus, some other features could be of interest. In order to compute stress distributions and deformations exactly, the Poisson's ratio is necessary. Furthermore, from this ratio, together with the Young's modulus, shear and bulk moduli can be derived. In some cases the weight could play a role. Therefore, the density should be known. Sometimes hardness values are quoted. Such data as far as being available will be included in the following for the sake of completeness.

II. MECHANICAL PROPERTIES

In this part all those mechanical properties mentioned before will be treated which are not directly strength related, but can either influence the reliability of a real structure by their influence on the stress distributions, e.g., as Young's modulus and Poisson's ratio, or correlate with the elastic and strength properties of ceramic materials, e.g., as density and porosity. Furthermore, hardness will be touched which can give some idea of the portion of plasticity in the material behavior.

At first the usual methods of measurements will be shortly described and their signifying value will be discussed. Then the range of data for the different materials will be shown which results from the mean values given in the literature. Finally some influences on and relations between the properties will be indicated.

A. Methods of Measurement

1. Young's Modulus

The most usual method to determine Young's moduli, E, is the measurement of load and deflection in a three- or four-point bending test with bars.[59,60] This procedure is the simplest and most suitable one for ceramics because it avoids some difficulties arising from the clamping of the specimens which are necessary for other commonly used quasi-static tests. Therefore, this method is recommended in most standard specifications for ceramics.[61-63] However, it has to be mentioned that the method exhibits some restrictions if higher accuracy is required. First of all, the bars have to be "long", i.e., the thickness and height have to be small compared with its length; otherwise the simple solution which is assumed for the stress distribution will become insufficient. Mostly a length to height ratio, which is the more important one, of at least 8 is demanded; in some cases, e.g., as for high anisotropy, this ratio should increase up to 24.[64] With respect to these demands the length specified for the standardization of bioceramics[62,63] seems to be too short. Furthermore, for low-modulus materials, additional mistakes can arise from the support conditions; their shape and some friction can influence the true length, deflection, and curvature.[65] On the other hand for high-modulus materials, the accuracy of the deflection measurement has to be very high because of the low deformations.

The last point is the reason that different dynamic test techniques are often used especially for dense high-modulus ceramics. These techniques measure either the propagation velocity of longitudinal ultrasonic waves in suitable specimens or resonance vibrations of bars which both depend on the Young's modulus.[59,66,67] The measurements can be performed very accurately and lead to reliable data because ceramics usually do not show any rate sensitivity. Difficulties can occur, however, at increased porosity or distinct viscoelasticity.

Using these dynamic methods further, elastic moduli, as bulk or shear modulus, can be determined, which can also be derived from the Young's modulus by simple formulas knowing the Poisson's ratio.[59,66] For porous materials, all these methods yield only to effective or apparent moduli, of course.

2. Poisson's Ratio

The direct measurement of the Poisson's ratio, ν, according to its definition by determining the transverse strain in an uniaxial tension test, will be mostly impossible for ceramics because of experimental difficulties and inaccuracies. The only quasistatic method which exists seems to be a bending test of rectangular thin plates and the measurement of the resulting biaxial deformation with high resolution which can be achieved by optical interference techniques.[59] The more usual methods, however, are the dynamic ones mentioned before. By measuring, in addition to the longitudinal wave velocity or resonance vibration, a further velocity or vibration, e.g., the shear wave velocity or the flexural vibration, respectively, which depend on the Young's modulus as well as on the Poisson's ratio, v can be determined.[66,67]

3. Density

The apparent or bulk density, ρ, which is mainly of interest, can be determined simply from the volume of a regular shaped body and its mass or weight, respectively ($\rho = M/V$). This is usually sufficient, unless a high accuracy is desired for the dynamic determination of elastic moduli for which the density has to be known. For this purpose, various techniques can be applied, such as pycnometer, buoyancy method, or sink-flat comparator,[59] which are all standard tests.[68] In case of porous materials, care has to be taken to seal the surface by a thin layer because all these methods use liquid environments.

4. Porosity

The porosity expresses, as a percentage, the ratio of the volume of the pores, V_p, to the exterior volume of a specimen, V. This definition, however, can have two meanings. The true porosity, P, includes all pores, open or closed, and can only be calculated from the density of the theoretically dense material ($P = 1 - \rho/\rho_o$). The apparent or open porosity, P★, describes only the portion of the open pores. It can be determined by filling the pores with some liquid (usually water or mercury) and measuring their volume from the mass difference between filled and dry specimen. Several standard methods recommend these procedures[68]. However, it has to be mentioned that especially in the case of small pores or low porisity because of capillary action, these methods can result in too low values.

5. Microhardness

Hardness definitions use the irreversible indentation in the material's surface caused by variously shaped bodies under defined loads and thus describe mainly the portion of plasticity in the material behavior. This portion is mostly very small for ceramics. Furthermore, because of their brittleness, they tend to crack which must be avoided corresponding to the hardness definition. Therefore, measurements with very small pyramidal diamonds[69,72] which generate high, but locally limited stresses, are preferred to determine the microhardness.[69] However, this method is also not very reliable. Although usually low loads are applied, small cracks often cannot be prevented.[70] These low loads create, moreover, only small, inaccurately measurable indentations which can be in the range of the grain size, so that additionally the assumption of a homogeneous material will not be fulfilled.[71] From this point of view, hardness data for ceramics have to be interpreted cautiously.

These values are also not very expressive for the mechanical behavior of ceramics. Although some purely empirical correlations to further mechanical properties seem to exist,[72] they are more useful to estimate "machinability" and wear behavior.

B. Range of Data

The data for these mechanical properties of the considered materials are summarized in Tables 1 to 3 as far as information available from the literature. These data result mostly from measurements with biomaterials, i.e., from materials which really have been used or developed for biomedical applications. In those cases where some material has been investigated only with respect to its biocompatibility and no mechanical data are mentioned, the authors tried to find some comparable data in the normal material science literature to show

Table 1
MECHANICAL PROPERTIES AND STRENGTHS OF CARBON MATERIALS

Material	Porosity (%)	Density (mg/m³)	Young's modulus (GPa)	Poisson's ratio	Microhardness (GPa)	Compressive strenth (MPa)	Tensile strength (MPa)	Flexural strength (MPa)	Fatigue strength (%)	Ref.
Graphite, isotropic or slightly anisotropic	7	1.8	25	—	—	—	—	140	70(d)ᵇ	73
	12ᵃ	1.8	20—24	≈0.2	—	65—95	24—30	45—55	50—60(d)ᵇ	74, 75
	16—20ᵃ	1.6—1.75	6—9	≈0.2	—	18—58	8—19	14—27	—	74, 81
	18	1.85	13.4	0.15	—	—	—	—	—	76
	31	1.55	7.1	0.25	—	—	—	—	—	76
	—	—	7.0	—	—	—	34	—	—	77
	—	0.1—0.5	—	—	—	2.5—30	—	—	—	78
Pyrolytic graphite, LTI	2.7	2.19	28—41	—	—	—	—	—	—	76
	—	1.3—2	17—28	—	—	900	200	340—520	—	77, 79, 80, 81, 85
	—	1.7—2.2	17—28	—	1.5—2.5	—	—	270—550	100	82, 134, 135
Vapor-deposited carbon	—	1.5—2.2	14—21	—	1.5—2.5	—	—	340—700	100	82
	—	—	9—14	—	—	—	30—230	—	—	77
Glassy (vitreous) carbon	—	1.4—1.6	—	0.28	1.5—2.0	—	—	70—205	100	82, 83, 84, 134, 135
	—	1.45—1.5	24—28	—	7	700	70—200	150—200	—	77, 85
	—	1.38—1.4	23—29	—	—	—	—	190—255	—	79, 86
	≤50ᵃ	<1.1	7—32	—	0.7—2.4	50—330	13—52	—	—	87

ᵃ Open.
ᵇ Dynamic (d).

Table 2
MECHANICAL PROPERTIES AND STRENGTHS OF BIOACTIVE AND BIODEGRADABLE MATERIALS

Material	Porosity (%)	Density (mg/m³)	Young's modulus (GPa)	Poisson's ratio	(Micro-hardness) (GPa)	Compressive strength (MPa)	Tensile strength (MPa)	Flexural strength (MPa)	Fatigue strength (%)	Ref.
Bioglass	—	—	—	—	—	—	56—83	—	—	43, 88
(-ceramics), glass ceramics	31—76	2.8	—	—	8—10	500	—	100—150	50(d)ᵃ	34, 38, 85
	—	0.65—1.86	2.2—21.8	—	—	—	—	4—35	—	89
Hydroxyapatite	0.1—3	3.05—3.15	7—13	—	4.2—4.5	350—450	38—48	100—120	—	36, 85, 90
	10	2.7	—	—	4.2	—	—	—	—	36
	30	—	—	—	—	120—170	—	—	—	91
	40	—	—	—	—	60—120	—	15—35	—	36, 53, 55
	2.8—19.4	2.55—3.07	44—88	—	—	310—510	—	60—115	—	119
	2.5—26.5	—	55—110	0.23—0.28	—	≤800	—	50—115	—	155
Tetracalcium-phosphate	"Dense"	3.1	—	—	—	120—200	—	—	—	51, 51, 85, 92
Tricalcium-phosphate	"Dense"	3.14	—	—	—	120	—	—	—	50, 51, 92
	36	—	—	—	—	7—21	5	—	—	54
Other calcium phosphates	"Dense"	2.8—3.1	—	—	—	70—170	—	—	—	50, 51
Mixtures	—	—	—	—	3.5	30—140	—	—	—	51, 93

ᵃ   Dynamic (d).

Table 3
MECHANICAL PROPERTIES AND STRENGTHS OF INERT CERAMICS

Material	Porosity (%)	Density (mg/m³)	Young's modulus (GPa)	Poisson's ratio	(Micro-hardness) (GPa)	Compressive strength (MPa)	Tensile strength (MPa)	Flexural strength (MPa)	Fatigue strength (%)	Ref.
Al₂O₃	≈0	3.93—3.95	380—400	0.26—0.32	24	4000—5000	350	400—560	70[b],60[d]	16, 17, 22, 94—97, 136
	25	2.8—3.0	150			500		70		98, 137
	35					200		55	84[b,c]	99, 137
	50—75	—	—			80				89, 137
MgO	≤5[a]	3.5—3.6	290—320	—	6—8	110	—	6—11.4	—	100—103
	25[a]	2.5—2.7	—	—				50—580		98
								40		
ZrO₂, stabilized	0[a]	4.9—5.56	150—190	0.36	5	1750	—	150—700	—	98, 101, 103, 104
	1.5	5.75	210—240			—		280—450		162
	5	—	150—200			—		50—500		21
SiC	28(25[a])	3.9—4.1	—			<400		50—65		12, 98, 137
Si infiltrated	0[a]	3.1—3.2	385—420	0.2	24	1500—3000	—	170—800		2, 105, 120
	0[a]	2.95—3.05	170—350	0.2—0.25	14	1200	180—(490)	300—360		22, 106, 107, 108, 120
Si₃N₄, hot pressed, reaction bonded	0[a]	3.15—3.2	300—330	0.26—0.3	15—22	2500—3000	300—(830)	600—1000	—	2, 15, 22, 105, 109, 110, 120
	20—35	2—2.6	80—210	0.24—0.26	4—7	600—1000	80—(355)	100—300		15, 22, 111, 112, 120
										120
SIALON, hot pressed, reaction sintered	—	3.2—3.25	300	0.28				220—830	—	113, 114, 115, 127
										164
	2—3	3.12—3.18	—					415—800	—	114, 116, 165
Ca-aluminate	9	2.93				—		340		113
	65(50[a])	—				—		11	50[b]	117
Ca-phosphate/Mg-aluminate	1.2	2.78				52		—		118

Note: Static, s; dynamic, d; corrosion, c.

[a] Open.

[b] Related to a specimen volume of 1 mm³ (m).

at least the order of magnitude. As far as possible, values or ranges of values are given for different porosities or densities, respectively, which influence properties and strengths most distinctly.

The data for the different materials and groups vary over a wide range. The "lightest" materials are the carbons. They are subdivided in Table 1 in four groups which are mainly characterized by the way of processing. This classification is unfortunate, but usual, and a consequence of the structural complexity of poorly organized carbons which does not seem to allow to describe them shortly by structural parameters.[82] These carbons exhibit densities around 2 mg/m³, except for the very porous ones, and approximate sometimes the theoretical limiting value of 2.2 mg/m³. The Young's moduli are low, comparable with those of plastics, and the hardnesses show the smallest values compared with the other ceramic materials. All these properties are very similar for the four classifications.

The bioactive and biodegradable materials, shown in Table 2, have densities of about 3 mg/m³ in the low-porous modifications. The Young's moduli seem to be even lower than those of the carbons, whereas the hardnesses are two to four times higher.

In contrast to this, the inert ceramics are strongly distinguished (Table 3). The densities are usually still greater, at least for the dense modifications, and can reach values over 5 mg/m³. However, the more significant difference lies in their extremely high stiffness. The Young's moduli are one order of magnitude higher than those of the other ceramics and are nearly twice as high as the normally used metals and alloys. A similar difference is observed for the hardnesses.

C. Relations Between Properties

From the data given in Tables 1 to 3 it is evident that strong correlations or dependencies between the different mechanical properties exist. Some of them are known or can be described by more or less empirical formulas, at least over a certain range.

The most obvious relation is that between density and porosity, P, given by the equation of definition (compare Section II.A.4):

$$P = 1 - \rho/\rho_o \tag{1}$$

where P is defined as a fraction and ρ_o means the density of the theoretically dense material. A more complicated correlation seems to exist between Young's modulus, E, and P or ρ, respectively. Usually it can be approximated by an exponential function:

$$E = E_o \cdot e^{-bP} \tag{2}$$

where E_o is the Young's modulus of the theoretically dense material and b is an empirical constant.[122] This expression can be converted with Equation 1 into

$$E = a \cdot e^{c\rho} \tag{3}$$

with $a = E_o e^{-b}$ and $c = b/\rho_o$. The different constants can be determined by a straight line fit of the data in a semilogarithmic plot[89,123] or by at least two measurements at different porosities or densities, respectively. For smaller variations of P or ρ, the exponential functions may be approximated by linear expressions.[89]

$$E = E_o \cdot (1 - bP) \tag{4}$$

$$E = a \cdot (1 + c\rho) \tag{5}$$

Equation 5 is sometimes written as[120]

$$E = d \cdot (\rho - \rho_1) + E_1 \quad \rho > \rho_1 \tag{6}$$

where E_1 is the Young's modulus for the density, ρ_1, and d again is an empirical constant $(d = a \cdot c)$. It must be regarded that E may additionally depend on the grain size, too. [79,81,124] Furthermore inhomogeneous porosity and pore shape anisotropy can influence E strongly. [125,126]

A dependence of the Poisson's ratio is known to the authors only in one case. For hydroxyapatite, ν decreases with increasing porosity. [155] It must be assumed that a similar effect should also exist for the other materials which possibly lies, however, within the accuracy of measurement.

Numerous investigations have been performed to determine some correlations between hardness and P, ρ, E, or grain size and various dependencies have been established, mostly empirical and restricted to certain materials and ranges of properties. For more detailed information, one must be referred to the special literature. [69-72.]

III. CONVENTIONAL STRENGTH PROPERTIES

Most strength data available at present for the ceramics considered are stress-related mean values. Further strain-related data are mentioned seldom because ceramics are usually extremely brittle.

These data are determined by different techniques and with varying specimen geometries. Because the failure of a specimen does not depend only on the material properties alone, but also on the size, shape, and distribution of flaws which unavoidably arise during the processing of the material and during the preparation, machining, and handling of the specimen, these varying conditions for measurement result in different strength values. These differences may become considerable and, therefore, it is necessary to know the test conditions in order to be able to interpret the data correctly and to transfer them to real structures for reliability estimations. In the following, the three mainly used test configurations and the data obtained with them will be discussed.

A. Methods of Measurement
1. Compressive Strength
The most favorite method appears to be the crushing test because it needs the simplest test equipment and specimen shape and possibly because it yields the highest strength values. Right cylindrical or rectangular bodies are compressed parallel to their axis up to failure and the strength is described by the monimal compressive stress at failure calculated from the load and the original sectional area. [128,129]

A material, however, can be separated only by shear or tension. Under the given boundary conditions, such stresses arise only in the interior of the body (whereas the surfaces are essentially under compression) and are lower than the nominal compressive stress, at maximum half of it in case of the shear stress. The stresses are additionally reduced, if the specimen height becomes comparable with its diameter, by friction in the interface between specimen and loading device.

Furthermore, failure will start always at any of the larger flaws and will be initiated at different stress levels depending the flaw size.* Since the internal flaws are usually smaller and less dangerous than those on the surface which are produced and deepened by machining

---

* In terms of fracture mechanics, if $\sigma \cdot \sqrt{a} \cdot f(geom) = K_{ic}$ where σ means the local stress is some characteristic length of the flaw, f(geom) takes into account the shape and position of it and K_{ic} is the fracture toughness of the material for normal (i = I) or shear (i = II, III) loading conditions, respectively.

and handling, much higher loads are necessary to destroy such a specimen by compression than, for example, by tension. Therefore, the fictively calculated "compressive strength" must result in much higher values than those obtained by arrangements which produce directly tensile stresses in the specimen.

Comparison of the different strengths in Tables 1 to 3 shows that the compressive strength for any material is at least three times and in some cases up to ten times higher than the according flexural or tensile strength. This is mainly due to the special loading and flaw conditions mentioned above, but may also result from the often used, too "short" specimens which influence strongly the measurement and let this fictive strength value increase significantly.

These considerations justify some doubts in applying these compressive strength values to the presumable behavior of a real structure. Even if the structure will be designed to bear mainly compressive forces as is usually proposed in the case of ceramic materials regarding the comparably high compressive strength, the stress field will be usually more complicated than in the simple specimen geometry and will exhibit tension components, which may — although smaller — confine the reliability of the structure.

2. Tensile Strength

Tensile strengths are determined under uniaxial tension by rupturing plane plates or cylindrical rods which are usually tapered in the central part. The strength is defined as the maximum stress in the smallest cross section at failure. This test is not very frequently applied to ceramics since considerable experimental difficulties arise from clamping the brittle specimens which require a very careful and expensive specimen preparation and a complicated set-up. Therefore, a standardized test exists only for carbon materials.[130]

This procedure, however, is the most critical one of all conventional techniques and yields the most conservative strength values. This is due to the fact that the whole specimen is subjected to tension. By this also the flaws at the surface which are often the greater and more dangerous ones in ceramics are exposed to this stress field. Since, furthermore, nearly the whole surface or volume, respectively, is stressed by approximately equal tension, the probability is very high that the greatest flaws which occur in the natural flaw distribution of the specimen are situated in this field so that the failure can start at a relatively low stress level according to the flaw size dependence of initiation.

These are the main reasons that the uniaxial tensile test results in the lowest strength values which can be obtained. It explains additionally, some influences of the specimen geometry.[97,131] In very small specimens, the existence of greater flaws is less probable than in big ones, provided that they are equally prepared and handled, so that the strength value can decrease with increasing specimen size. This has to be taken into account for structural problems if the high stressed areas in the structure differ significantly from those considered in the specimen. In any case, the tensile strength would be the most reliable conventional value to estimate the structural safety if the stress distributions are known.

3. Flexural Strength

The more frequently applied method to determine strengths for tensile stresses is the bending test because of the minor experimental difficulties. It is performed in the same arrangement as used for the measurement of Young's moduli[59,60] and also is recommended as the standard test.[61-63] The so-called bending or flexural strength is defined as the maximum stress on the tension side of the bar at failure. This stress is calculated assuming the simple beam theory.[64]

The values determined in this way are usually higher than the comparable tensile strengths; they can amount up to twice as much especially for ceramic materials (compare Tables 1 to 3). The reason is essentially the same as for the specimen size influence in the pure tensile to high tensile stresses. Therefore, the probability for the greater flaws to be situated just in this part is small, and higher stress levels are necessary to initiate failure at one of the remaining smaller flaws. Similar to the tensile test, an effect of specimen size can also be observed, of course,[131,132] and additionally an influence of the bending arrangement is observed if going from three- to four-point bending.[120,133] In the latter case the high-stressed region increases likewise which explains the lower flexural strengths measured usually in four-point bend tests.

A further influence of the arrangement will be observed if the specimens are wrong-shaped, i.e., if the bars are too short compared with their height (compare Section II.A.1). Then the simple bending theory will no longer be valid, the stress distributions become more complicated, and the calculated strengths assuming the simple distributions increase.[131]

All these effects require a critical examination of strength values if they should be applied to structures. Stress distributions and stressed regions have to be compared and size effects have to be considered. This can be done if it is presupposed that specimens and structure are processed and prepared under equal conditions. Otherwise, additional effects can arise from differences in the statistical flaw distributions.

4. Fatigue Strength

Most materials show a time-dependent failure behavior or, in other words, a decreasing strength with increasing loading time. This is true also for ceramics. The process of failure in such brittle materials, however, is completely different from that in ductile materials. Nevertheless, the term "fatigue", which described originally the behavior of metals under cyclic loads and is mainly determined by their plasticity, is used for ceramics, too. This may not be misunderstood and is meant only in the sense of time-dependence. Furthermore, in the case of ceramics, this behavior is distinguished by two additional, eventually misleading terms. Whereas the failure under alternating, mostly cyclic loads is called "dynamic fatigue", similar to other materials groups, the behavior under long-term constant load which is similar to creep is named "static fatigue". Both "fatigue" strengths are usually determined in the bending arrangement by applying either a constant load or a cyclic load with constant amplitude and by measuring the time (or number of cycles) to failure.

The decrease in strength which is reported for biomedical ceramics, amounts up to 50%,[34,75,96,117] however, sometimes no reduction seems to be found[82,134] (compare Table 1 to 3). Normally, also a huge scatter is observed in the data. Therefore, it appears to be doubtful that this procedure is useful for ceramic materials, and improved methods which will be described later should be applied to characterize the long-term behavior.

B. Range of Data

An evaluation of the different materials with respect to their strengths appears to be difficult because the data vary strongly, not only between and in the groups, but also for each material itself (compare Table 1 to 3). Even if the porosity or density remains the same, large variations are observed. This means the strength is determined by further influences like the properties of the basic materials and the way of processing the ceramic. Therefore, only a rough comparison can be made from the data available. Nevertheless, looking, for example, at the modifications with the lowest porosity, it becomes obvious that the bioactive and the biodegradable materials exhibit the lowest strengths (Table 2). The graphites are comparable to them, whereas the further carbon materials show significantly higher strengths and can reach values comparable with those of some inert ceramics. The latter ones are certainly the materials by which the highest strengths can be obtained. This is true for alumina, the best-investigated material of this group, but even more especially for the silicon-based materials which could offer further improvements with respect to strength if their compatibility would be verified. By these ceramics tensile strengths could be attained which are in the range of the metallic biomaterials used today.

C. Influences on Properties

Besides the influences of test method, specimen size, and flaw distribution due to prepaaration which have been discussed before (see Section III.A), the strength of a ceramic depends extremely on the porosity or density, respectively. As can be seen from Tables 1 to 3 and as shown by several authors for defined material compositions,[36,80,81,89,120,137] the strength decreases strongly with increasing porosity or decreasing density, respectively. The decrease can amount up to two orders of magnitude if going, for instance, from 0 to 50% porosity.[120,137] This strength-porosity dependence can be described over a wide range by the following expression:[138]

$$\sigma_p = \sigma_o \, e^{-BP} \tag{7}$$

where σ_p and σ_o mean the strengths of the porous and the dense material, p is the porosity, and B is an empirical constant resulting from the slope of the $\ln\sigma_p$ vs. P plot. Applying Equation 1, this expression can be transferred into a strength-density dependence just as in the case of the Young's modulus (Equation 3). For small porosity or density ranges, the two expressions may be approximated again by linear functions similar to those found in Section II.C.[127] On the other hand, if the porosity variations become too large (more than 40 to 50%), the exponential description seems to become insufficient[137] and a more complicated relation has to be considered.[138] Looking in more detail on the influence of different pores, additionally a slight dependence on the pore shape and orientation can be observed.[137]

A further strong variation of the strength is caused by different grain sizes. At constant porosity, usually the strength decreases with increasing mean grain sizes.[16,124,139-142] This behavior can be described mostly with sufficient accuracy by the Knudsen equation:[141]

$$\sigma_g = \sigma_1 \, G^{-\alpha} \tag{8}$$

where σ_g is the strength for the mean grain size G, and σ_1 and α are constants (depending on porosity and temperature and with α between 0.3 and 0.5). This equation becomes equal to that proposed by Orowan if $\alpha = 0.5$ which appears to be valid especially in the range of larger grain sizes.[142] For small grain sizes, an extended Orowan equation is advantageous which was introduced by Petch.[142]

$$\sigma_g = \sigma_\infty + \sigma_1 \cdot G^{-1/2} \tag{9}$$

with an additional constant $\sigma\infty$. If considering a large grain size range, a combined Orowan-Petch treatment with two branches can be useful which will be more accurate than Equation 8.[142]

Because of the power function character of the strength-grain size dependence, it is evident that remarkable improvements in strength can be achieved by reducing the grain size. This is especially important in the case of the dense inert ceramics where the relatively high strength obtained by low porosity can be further increased. For alumina, for example, the flexural strength increases by a factor of 3 to 4 going from 100 to 10 μm[140,141] and by another factor of about 2 between 10 and 2 μm[124,140] which is the grain size range of the bioceramics used today.

The two influences of porosity and grain size cannot be separated exactly in all cases because both are often correlated, especially for the reaction-sintered materials. Therefore, a simple combination of Equations 7 and 8 by[141]

$$\sigma = k \cdot e^{-BP} \cdot G^{-\alpha} \tag{10}$$

can only be an approximation because at least α depends additionally on P.[140]

These strength dependencies can be explained, at least partly, by varying flaws, because the distribution of the possible flaws is certainly determined by the size of pores and grains. The correlation, however, is not quite as simple since the initial flaw distribution can obviously vary for different manufacturing conditions and can be additionally changed by finishing processes as polishing, firing, or quenching which usually increase the strength* significantly.[143]

Further influences on strength are given by microstructural aspects, of course. This becomes evident in the case of impurities which normally reduce it.[124,144] Nevertheless, admixtures could be advantageous for the long-term behavior because they sometimes enlarge the fracture toughness[144] (compare Section IV.A).

Finally, environmental effects can lead to changes in strength. For biomedical applications, the influence of water or aqueous solutions is interesting. They appear to cause some decrease even under quasi-static conditions.[145,155] More importantly however, are the reductions under long-term conditions. Aging without loading as well as static loading for a long time in aqueous environment usually diminishes the strength remarkably.[11,12,43,54,99,117,146,155] Sometimes only a slight increase is observed[146] which could be explained by rounding the flaw or crack tips and, thus, mitigating them.

D. Statistical Treatment

Every measurement of a physical quantity — even if performed under apparently constant conditions — shows a scattering of the results. This can be caused by two reasons: firstly, every experimental set-up exhibits unavoidable inaccuracies, for example, due to friction, temperature effects, etc., and secondly, the quantity itself which is to be measured scatters in different specimens.

The strength of ceramic materials is mainly subject to the second type of scattering; compared with it, the first type can be neglected. This results in the well-known finding that one has to accept SDs of 25% as not unusual in strength measurements, or the data of nominally identical specimens tested under nominally identical conditions can range from one third to three times their mean value. Hence, the significance of the mean value for purposes of material evaluation and design is rather dubious. Therefore, it is obviously necessary to consider the scatter of the test results as an integral part of the phenomenon.

This scatter of ceramics is due to the fact that the failure by brittle fracture or the brittle fracture strength of a material is related to its defect structure, i.e., to the concentration and severity of defects in the specimen. Because the failure is always initiated at the most dangerous defect which exists, at a stress level which depends on the defect geometry, and because, on the other hand, shape and size of the defects vary strongly in ceramics, large differences in the rupturing stresses may be observed for different specimens.

Whereas the effect of measurable defects can be taken into account by fracture mechanics (compare Section IV.), small defects will be considered mostly by statistical methods. This is frequently done by assuming a Gaussian distribution for the strength data and calculating the SD. This should be sufficient if large numbers of specimens with small scatter are treated. From the theoretical point of view, however, the application of a symmetrical distribution which allows strength values below zero is not correct. Therefore, asymmetrical extreme value distributions are preferred today.

The most usual one is the Weibull distribution.[147,148] It has been derived using the following assumptions:

1.   The fracture of the bulk specimen is determined by the local strength of its weakest

---

* The strengthening effect by these finishing processes is often caused not only by minimizing the flaws, but also by inducing a superficial compression layer which reduces the tensile stresses generated under load by superposition.

volume element, similar to the rupture of a chain at its weakest link (''weakest-link hypothesis'').

2.   Failure cannot occur below zero-applied stress.
3.   Failure must occur for sufficiently large stresses.

This leads to the probability for failure.[120]

$$F = - \exp\left[- \int_V \left(\frac{\sigma - \sigma_u}{\sigma_o}\right)^m \cdot dv \right] \quad \sigma > \sigma_u$$
$$F = 0 \qquad\qquad\qquad\qquad\qquad\qquad \sigma < \sigma_u \tag{11}$$

where σ means the stress field in the loaded component which can vary spatially, and σ_u, σ_o, and m are constants. The three parameters of the distribution are defined as follows:

1. σ_u Is the threshold stress below which the failure probability is zero.
2.   m Is the Weibull modulus and a measure of the variability of the quantity ($\sigma - \sigma_u$) if σ is the fracture stress.
3. σ_o Is the third material parameter.

According to Equation 11, F depends on the stress distribution and the size of the component or specimen considered. For simple tension, for instance, σ is uniform and the equation becomes

$$F = 1 - \exp\left[- \left(\frac{\sigma - \sigma_u}{\sigma_o}\right)^m \cdot V \right] \quad \sigma > \sigma_u$$
$$F = 0 \qquad\qquad\qquad\qquad\qquad\qquad \sigma < \sigma_u \tag{12}$$

For more complicated distributions and geometries, it could become difficult to solve the integral. Then, often Equation 11 can be approximated by Equation 12, assuming an effective volume, V_{eff}, which describes the highly stressed, most endangered part of the total volume.[97]

In practice, best estimates of σ_n are usually relatively small compared to the mean strength. In view of the dangers of overestimating σ_u, it should be set equal to zero unless there are physical reasons to expect an upper limit for the size of the flaws. Thus, a distribution results which is determined by only two parameters, σ_o and m, and σ_o gets the physical meaning that a specimen of unit volume subjected to a stress σ_o has a failure probability of $1 - 1/e$ or 0.632.

Following Equation 12, the strength dependence on specimen size, which has been mentioned repeatedly, can clearly be shown. The ratio of the strengths, σ_i, for two different sizes, 1 and 2, are connected with their volumes V_i, or effective volumes, respectively, by

$$\frac{\sigma_1}{\sigma_2} = \left(\frac{V_2}{V_1}\right)^{1/m} \tag{13}$$

This relation is used, for example, also to normalize the strength values to a unit volume (compare Table 3).[106,109,111] It is proved by many investigations.[86,131] Difficulties can arise, however, if different flaw distributions exist, e.g., due to surface flaws and volume defects. If only surface flaws are responsible for failure, it should be possible to replace the volume integral in Equation 11 by a surface integral.[147-150] Otherwise, more refined models have to be taken into account.[149]

Normally it is advantageous to consider the cumulative distribution function and to plot it, depending on the fracture stresses, σ_j of the different specimens in a particular logarithmic diagram by[150,151]

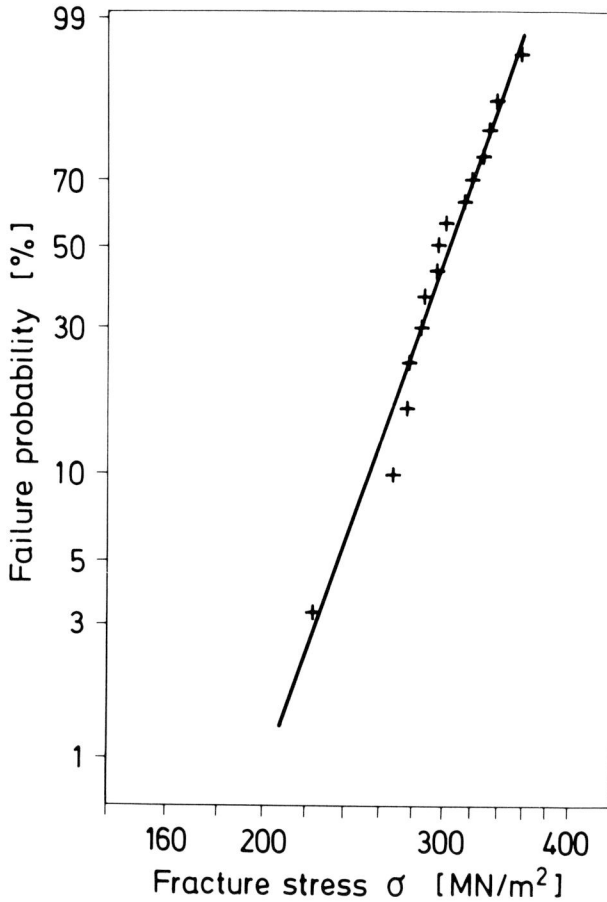

FIGURE 1.   Weibull diagram for a technical alumina tested in three-point bending; regression analysis leads to m = 10.7 and σ_o = 312.6 MN/m$_2$.

$$\ln \ln \frac{1}{1 - F} = m \cdot \ln \left(\frac{\sigma_j - \sigma_u}{\sigma_{BO}} \right)$$ (14)

where additionally σ_o is replaced by $\sigma_{BO} \cdot V_{eff}{}^{1/m}$ (compare Equation 12). Thus, the curve becomes a straight line with the slope m, which is demonstrated by an example in Figure 1. In this representation the failure probability for a given stress level can easily be read. It becomes evident, furthermore, that m describes the variability of the strength data — a steep slope or a large m means small scatter and a small m means large data variations — and that σ_{BO} determines the position of the curve in the diagram similar to a mean strength.

Tables 4 to 6 show the Weibull parameters for the material groups considered and as far as available, assuming a two-parametric distribution (according to Equation 11) with σ_u = 0), and derived mainly from bending tests. The material parameters, σ_o — according to their similarity to a strength value — show the same tendencies as the mean strengths in Tables 1 to 3. The Weibull moduli are relatively high which means small scatter.

IV. FRACTURE MECHANICS APPROACH

In Section III, it was emphasized repeatedly that the strength of a ceramic specimen depends strongly on the occurrence of flaws which can be either extrinsic, e.g., due to

Table 4
STATISTICAL AND FRACTURE MECHANICS PARAMETERS OF CARBON MATERIALS

Material	Designation/ characteristics	Environment/ test condition	Weibull parameters		Fracture toughness, K_{Ic} [MNm$^{-3/2}$]	Crack growth parameters[a]			Ref.
			σ_{BO}	m		n	^{10}log A	Method[b]	
Graphite	P = 18%	Air	—	—	0.73	113	16	dt	76
	P = 31%	Air	—	—	1.46	218	−35	dt	76
Pyrolytic graphite	P = 2.7%	Air	—	—	2.5	216	−88	dt	76
Glassy (vitreous) carbon		Air	—	—	0.71 (−1.08)	178	31	dt	8+
	CK 1 + 2	Air	—	—		78—82	10—13	dt	152
	Material I	Air	90—120		1.02—1.06	57—154	40—129	dt	83
		Water							
	Material II	Air	115—145		0.81—1.28	59—92	35—62	dt	83
		Water							

a   For V in m/s and K in MNm$^{-3/2}$.
b   Double torsion test, dt.

Table 5
STATISTICAL AND FRACTURE MECHANICS PARAMETERS OF BIOACTIVE AND BIODEGRADABLE MATERIALS

Material	Designation/ characteristics	Environment/ test conditions	Weibull parameters		Fracture toughness K_{Ic}[MNm$^{-3/2}$]	Crack growth parameters[a]			Ref.
			σ_{BO}	m		n	^{10}log A	Method[b]	
Glass ceramics	Li—Al—Si	Air	—	—	—	32.6	—	df	153
	Li—Si	Air	—	—	1—3.5	—	—	—	154
Hydroxyapatite	Different porosities (2.5—26.5)	Dry	96—124	4.4—9.3	0.6—1.1	11; 26	−1.7	dt; df	155
		Wet	66—115	4.8—9.4	0.4—0.7	12	—	df	155

a   For v in m/s and K in MNm$^{-3/2}$.
b   Dynamic fatigue, df; Double torsion test, dt.

Table 6
STATISTICAL AND FRACTURE MECHANICS PARAMETERS OF INERT CERAMICS

Material	Designation/ characteristics	Environment/ test condition	Weibull parameters $\sigma_{BO}(\sigma_u)$	m	Fracture toughness, $K_{Ic}[MNm^{-3/2}]$	Crack growth parameters[a] n	^{10}log A	Method[b]	Ref.
Al$_2$O$_3$	—	—	—	—	—	21.5	—	df	156
	—	Air	—	—	—	40	—	dt	
	A/dense	Ringer's solution	—	—	—	56	-37	dt	152,157
	B/dense	Air	—	—	—	42	-27	dt	
		Ringer's solution	—	—	—	63	-42.5	dt	152,157
	C/dense	Air	—	—	—	71	-44.5	dt	
		Ringer's solution	—	—	—	54	-42	dt	152,157
			—	—	—	50	-38		
	Friialit/dense	Inert	335	—	—	46.1	—	df	158
		Ringer's solution	—	—	—	44.8	—		
	M9/large grains		—	—	—	35	—	df	159
	M8/small grains	Water/Ringer's solution/serum	380—440	—	—	68	—		
	5—7% porosity	Water	—	—	—	35.7	-24	—	160
		Saline solution	—	—	—	60.4	-30.6	dt	
		Medium 199	—	—	—	75.5	-42	—	
	FAO	Air	354	13.9	4.81	22.7	—	df	161
		Ringer's solution	299	15.8	4.06	22.6	—		
	A 975	Air	355	16.7	4.31	31.4	—	df	161
		Ringer's solution	316	15.3	3.67	38.2	—		
	AP 997	Air	349	13.3	4.92	45.7	—	df	161
		Ringer's solution	325	16.0	4.08	41.0	—		
ZrO$_2$	Partially stabilized 5%; 35% tetragon	Water	—	—	3.5—4.6	80/51	—	dt/df	162
			—	—	3.5—7	50/61	—		
SiC	Self-bonded	—	—	8	4.4—5	—	—	—	120,121

Table 6 (continued)
STATISTICAL AND FRACTURE MECHANICS PARAMETERS OF INERT CERAMICS

Material	Designation/ characteristics	Environment/ test condition	Weibull parameters		Fracture toughness, $K_{Ic}[MNm^{-3/2}]$	Crack growth parameters[a]			Ref.
			$\sigma_{Bo}(\sigma_u)$	m		n	^{10}log A	Method[b]	
Si_3N_4	Hot pressed	—	325(90)	15	2.6—5.7	—	—	—	120,121
	Si–infiltrated	—	300—490	≥10	4.4—5	100	—	sf	106,107,156
	Hot pressed	—	500—830	6—20	2.8—8	—	—	—	109,110,120,121
	Reaction bonded	—	100—350	10—22	2—4.1	70	—	df	111,112,120,121,156
	Reaction bonded	Air, water saturated	—	11.8	—	56.8	—	sf	163
SIALON	Hot pressed	—	—	—	5—6	—	—	—	164,165
	Sintered	—	—	10	6	—	—	—	165

[a] For v in m/s and K in MNN^{-32}.
[b] Double torsion test, dt; dynamic fatigue, df; static fatigue, sf.

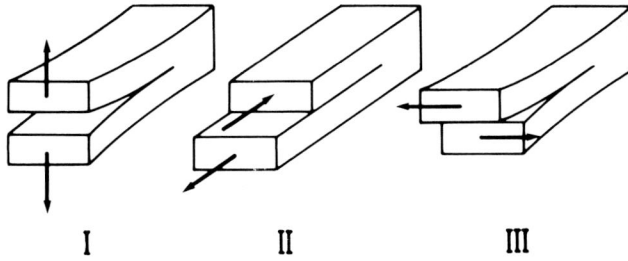

FIGURE 2. Basic modes of fracture. (I) Opening; (II) first shear or (edge) sliding; (III) second shear or tearing mode.

machining or surface finishing processes, or intrinsic, e.g., due to the material production process. It is in principle not possible to measure the flaw independent strength properties of a material by conventional strength tests.

In contrast to such tests, in the fracture mechanics approach, the influence of flaws or cracks on the strength behavior is taken into account. Consequently, in fracture mechanics experiments, sharp cracks of definite geometrical extension are used as a probe to establish the resistance of the material to crack propagation. It will be shown, furthermore, that by using fracture mechanics parameters also, the influence of time-dependent stressing conditions and of environmental media on the strength behavior can be quantitatively measured.

A. Basic Concept

1. Stress Intensity Factor

The most important fracture mechanics parameter is the "stress intensity factor", K. Actually three different K-factors have to be considered because it has been proven useful to distinguish three basic "modes" of crack-surface displacement, as shown in Figure 2. K_I refers to an applied tensile stress perpendicular to the crack walls, K_{II} refers to shearing of the crack walls in the direction of crack motion, and K_{III} refers to mutual shearing perpendicular to the direction of crack motion. By superposition of these three modes, any arbitrary loading situation of a crack can be described.[166] For brittle materials, as ceramics and glasses, the "crack opening" mode I, is by far the most important one. Therefore, in the following, only K_I values will be considered. For this loading type, the stresses in the neighborhood of the crack tip ($r \ll a$) in a plane perpendicular to the crack front are given by

$$\sigma_{xx} = \frac{K_I}{\sqrt{2\pi r}} f_{xx}(\vartheta)$$

$$\sigma_{yy} = \frac{K_I}{\sqrt{2\pi r}} f_{yy}(\vartheta)$$

$$\sigma_{xy} = \frac{K_I}{\sqrt{2\pi r}} f_{xy}(\vartheta) \tag{15}$$

where a, r, and ϑ are defined in Figure 3. This means that the stresses increase with approach to the crack tip ($r \to 0$) — theoretically up to infinity, in reality they are confined by plastic effects — and that the stress intensity factor K_I determines this increase.

In ceramics the cracks are usually very small compared with the dimensions of the specimen or component, respectively. Therefore, the stress field which governs the crack (i.e., which would occur in this region if the crack would not be there) can be assumed to be uniform

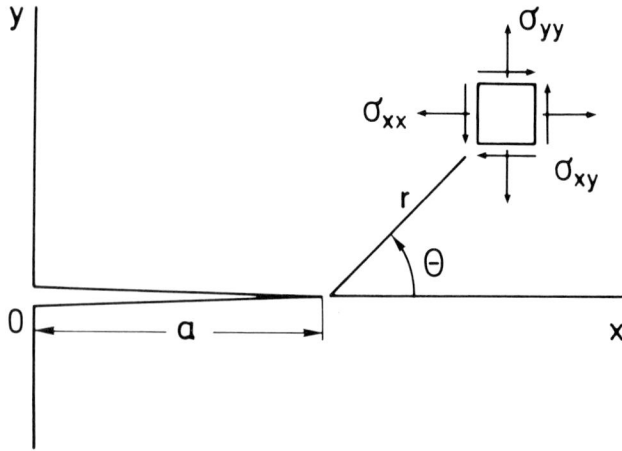

FIGURE 3.   Definition of stresses and coordinates at the crack tip.

and stress gradients can be neglected. In this case the stress intensity factor can be expressed by

$$K_I = \sigma \cdot \sqrt{a} \cdot f(a/d) \tag{16}$$

where $f(a/d)$ is only a function of the crack size, a, some characteric dimension, d, of the specimen and the geometry. It is evident from Equation 16 that K_I is a measure for the intensification of the applied stress due to the presence of a crack.

Because of $a \ll d$ and since also the range of the possible crack sizes is usually small compared with d in ceramics, the function $f(a/d)$ can be mostly replaced by a constant factor which is only determined by the geometry. For the materials considered, surface cracks are of special interest. In the case of a long shallow surface crack of the depth a, which can be regarded as a semielliptical crack having minor and major axes a and c, respectively, with $c \gg a$, the stress intensity factor becomes

$$K_I = 1.12 \, \sigma \cdot \sqrt{\pi a} = 1.990 \, \sigma \cdot \sqrt{a} \tag{17}$$

and for the semicircular surface crack ($c/a = 1$),

$$K_I = 1.212 \, \sigma \cdot \sqrt{a} \tag{18}$$

In comparison to that the K-values for similar internal cracks are

$$K_I = \sigma \cdot \sqrt{\pi \, a/2} = 1.253 \, \sigma \cdot \sqrt{a} \tag{19}$$

for the long narrow elliptical crack with the smaller axis $a/2$ or the width a, respectively; and

$$K_I = 2 \, \sigma \cdot \sqrt{a/2 \, \pi} = 0.798 \, \sigma \cdot \sqrt{a} \tag{20}$$

for the circular crack with the diameter, a. The Relations 17 to 20 demonstrate that surface cracks are more dangerous than internal ones, and, likewise, that narrow, laterally extended cracks are more critical than "penny-shaped" ones. Further extensive compilations of stress intensity factors are given in References 167 and 168.

2. Fracture Toughness

The use of the stress intensity factor approach to strength problems is based on the experimentally evidenced fact that catastrophic failure occurs when K_I reaches a critical value, K_{Ic}. This critical K_{Ic} is, within certain limits, independent of specimen size or type of loading and may therefore be considered as a real material parameter. It is a measure for the resistance of the material against fast crack extension and is called the "fracture toughness".

Because of the particular stress-crack size-dependence of K_I (compare Equation 16), it is evident that the critical value K_{Ic} will be reached at a large crack by a lower stress than at a small defect. This explains the dependence of strength on the flaw sizes, or evaluating different materials a low fracture toughness means that the material can tolerate only small cracks.

B. Slow Crack Growth

The fracture toughness, K_{Ic}, must not be regarded to be the only relevant parameter for characterizing the strength behavior of ceramics because it is well established today that also at K_I-values below K_{Ic} a crack can extend in ceramic materials. This crack extension can occur at very low velocities until the crack reaches a length which results in the critical combination of load and crack length, i.e., in K_{Ic}, where fast crack propagation occurs. In this way the time dependence of the strength of ceramic components under load can be understood as caused by slow crack growth. Hence, in order to estimate the susceptibility of a material to delayed failure or to predict the lifetime of a load-bearing ceramic component, it is not sufficient to know the fracture toughness alone. To evaluate time-dependent failure, one has to look additionally into the dependence of crack propagation on the load.

1. v-K_I-Dependence

The basic investigations into the dependence of crack velocity, v, on the load were performed observing the propagation of deliberately introduced macroscopic cracks. As a main result of all these investigations, it was found that — for defined environmental conditions— there exists a one-to-one dependence between crack velocity, v, and applied stress intensity factor, K_I. This appears to be true for all glasses and ceramics investigated up to now.

A typical example of such v-K_I-curves obtained for a particular alumina is shown in Figure 4, which demonstrates simultaneously the influence of environments. As it was indicated at first by Wiederhorn[170] for soda-lime glass, the crack growth curve at low crack velocities can be subdivided into three regions. In Region I crack extension is governed by the rate of stress corrosion at the crack tip. In Region II the crack velocity is much less dependent on the stress intensity and is assumed to be determined by the rate of diffusion of the corroding species to the crack tip. In Region III the crack velocity is independent of the surrounding medium, because the corroding substances cannot follow the crack tip; the curve runs into that one which would be measured under vacuum conditions.[171]

These v-K_I-curves can be analytically described by refined crack formation models.[172] Mostly, however, only the very slow velocity range of Region I is of interest, and for this part a relatively simple description has been proposed by Evans and Wiederhorn:[173]

$$v = A \cdot K_I^n \tag{21}$$

where A and n are purely empirical constants determined from a straight line fit of the experimental data in a log-log diagram (compare Figure 4). This power function form is well established meantime and appears to be of sufficient accuracy.

C. Lifetime Calculations

The major aim of studying slow crack growth is to develop a basis for failure prediction.

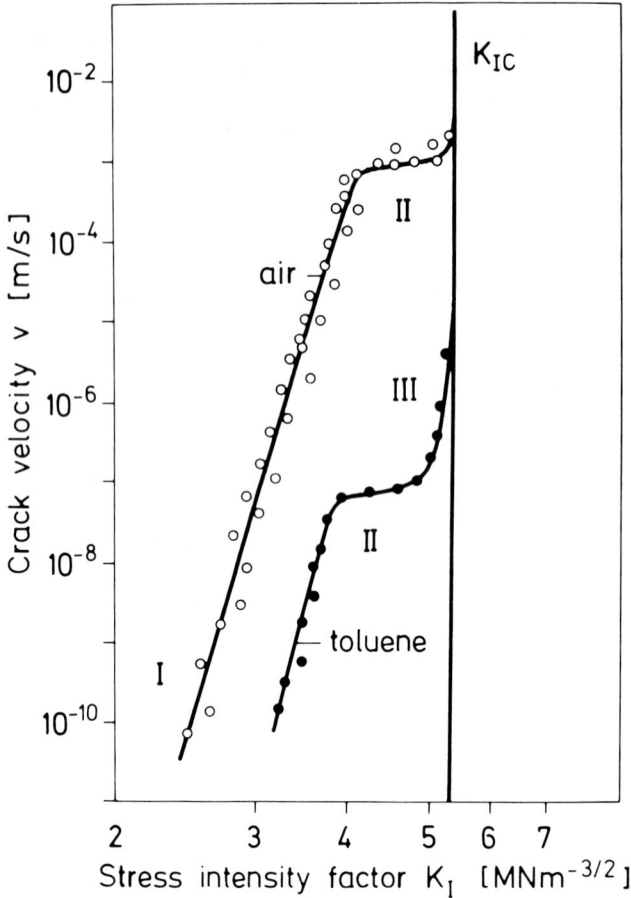

FIGURE 4.   v-K_I-dependence for an alumina ceramic in moist air (50% relative humidity) and toluene.[167]

The guiding idea in this fracture mechanics-based approach is to calculate the time a crack needs for increasing by slow crack growth to the critical length for a given load.[174] This can principally be done if v is a unique function of K, by solving the differential equation da/dt = v(K) which defines the crack velocity. Separation and integration lead to[152]

$$\int_{o}^{t_f} dt = \int_{a_o}^{a_c} \frac{da}{v(K)} \tag{22}$$

where t_f means the time to failure or the lifetime, respectively, a_o is the initial crack length at the time t = 0, and a_c is the critical crack length at failure.

Assuming the simplifications which are valid for ceramics, i.e., a K-value which can be described by Equation 16 with f = constant (e.g., compare Equations 17 to 20) and a v-K dependence which can be expressed by a power function as in Equation 21,* the differential equation becomes

---

*   By this assumption, the possibly different parts of the curve for high velocities (compare Figure 4), as well as those for very slow ones (which cannot be measured), are described by Equation 21. This might not be quite correct. Fast crack propagation, however, contributes only a very small amount to the total lifetime which is usually desired to be long, so that errors in this amount can be neglected. In the very slow velocity range, one can imagine from physical reasons only a steeper slope of the curve or, in the extreme case, a lower crack growth limit in K which is supposed by some authors, but not generally proved.[173] This would lead to longer lifetimes than calculated and therefore, at least, to a conservative safety estimation.

$$\frac{da}{dt} = A \cdot f^n \cdot \sigma (t)^n \cdot a^{n/2} \tag{23}$$

where σ can exhibit any time dependence. In general this equation cannot be solved analytically in closed form, but numerical methods must be applied.[152] For some simple loading cases, however, analytical solutions exist which will be outlined in the following because they allow the estimation of lifetimes for similar conditions or are used to determine the material parameters.

1. Constant Load — Static Fatigue

The time to failure, t_s, under a constant tensile stress, σ_s, i.e., under static load, can easily be derived from Equation 23 by substituting σ_s for $\sigma(t)$ and integrating this differential equation by the Procedure 22:

$$t_s = \int_o^{t_s} dt = \frac{1}{A \cdot f^n \cdot \sigma_s^n} \int_{a_o}^{a_c} \frac{da}{a^{n/2}}$$

$$t_s = \frac{2}{A(n - 2 \cdot f^n \cdot \sigma_s^n}) [a_o^{(2-n)/2} - a_c^{(2-n)/2}] \tag{24}$$

By replacing the crack lengths a_o and a_c by the corresponding initial stress intensity factor, K_{Ii}, and the fracture toughness, K_{Ic}, respectively, using Equation 16, the relation results which is normally presented for t_s and which is obtained by a slightly different derivation:[169,173-175]

$$t_s = \frac{2}{A(n - 2) \cdot f^2 \cdot \sigma_s^2} (K_{Ii}^{2-n} - K_{Ic}^{2-n}) \tag{25}$$

Generally, it holds $K_{Ic}^{2-n} \ll K_{Ii}^{2-n}$ since normally $n > 10$ for ceramics and $K_{Ii} < 0.9 K_{Ic}$ in practical applications, so that to a good approximation Equation 25 becomes

$$t_s = \frac{2}{A(n - 2) \cdot f^2 \cdot \sigma_s^2} K_{Ii}^{2-n} \tag{26}$$

It is evident from Equations 24 to 26 that the time required for the initial flaw to grow to a dimension critical for catastrophic crack extension, i.e., the lifetime, decreases with increasing stress, σ_s, and with increasing initial flaw depth, a_o, or K-value K_{Ii}, respectively. This behavior is commonly known as "static fatigue".

Because the inert strength, σ_{in}, of a material, i.e., the strength determined under inert environmental conditions and with high loading rates so that no significant subcritical crack growth occurs prior to fracture, is correlated with K_{Ii} by

$$K_{Ic} = \sigma_{in} \cdot \sqrt{a} \cdot f \tag{27}$$

the initial K-value for constant load $K_{Ii} = \sigma_s \cdot \sqrt{a} \cdot f$ can also be expressed as[175]

$$K_{Ii} = \frac{\sigma_s}{\sigma_{in}} \cdot K_{Ic} \tag{28}$$

By substituting this relation and setting

$$B = \frac{2}{A(n-2) \cdot f^2 \cdot K_{Ic}^{n-2}} \tag{29}$$

Equation 26 is often written in the form[175]

$$t_s = B \cdot \frac{\sigma_{in}^{n-2}}{\sigma_a^n} \tag{30}$$

where B is again a material constant for a given environment.

2. Constant Loading Rate — Dynamic Fatigue

A similar dependence is obtained for increasing load which is interesting in particular to explain the influence of stressing rate on the fracture strength.[176] In the following only a constant stressing rate $\dot{\sigma}$ is considered, i.e., $\sigma(t) = \dot{\sigma} \cdot t$ is supposed. Hence, from Equations 22 and 23, it follows for the time to failure[152]

$$t_d = \left[\frac{2(n+1)}{A(n-2)} \cdot \frac{1}{f^n \cdot \dot{\sigma}^n \cdot a_o^{(n-2)/2}} \right]^{1/(n+1)} \tag{31}$$

neglecting again the term with a_c or K_{Ic}, respectively, as in Equation 26. Considering the fracture strength which is usually measured under these conditions and is determined by $\sigma^f = \dot{\sigma} \cdot t_d$, it results:[152,176]

$$\sigma_f = \left[\frac{2(n+1)}{A(n-2)} \cdot \frac{\dot{\sigma}}{f^n \cdot a_o^{(n-2)/2}} \right]^{1/(n+1)} \tag{32}*$$

or using the Relations 28 and 29,[175]

$$\sigma_f = [B(n+1) \cdot \sigma_{in}^{n-2} \cdot \dot{\sigma}]^{1/(n+1)} \tag{33}$$

From these relations it is seen that the fracture strength decreases with increasing stressing rate since the flaws are given more time to grow. This behavior is mostly called "dynamic fatigue" in the ceramic literature which is not very fortunate, comparing it with the totally different meaning of this designation for other material groups.

Equation 33 is of fundamental importance in determining the slow crack growth parameters A and n using unnotched specimens, i.e., specimens containing no deliberately introduced initial cracks.[175] In a similar way as for increasing load also the effect of a load decrease can be considered. This is, for example, important to estimate the additional crack growth during the unloading phase after proof testing.[152]

---

* This approximate relationship is mostly sufficient. In a more detailed consideration, however, also the effects which could arise from deviations of the v-K_I curve from the assumed power function can be taken into account.[176]

3. Alternating Loads — Cyclic Fatigue

In practice for service, alternating or cyclic loads are the most important ones. It is generally assumed that under these conditions the v-K_I-dependence is the same as for constant stress, i.e., that there is no significant crack propagation enhancing effect of cycling. From this assumption it can be derived that the failure time, t_c, for any periodic loading is directly proportional to the failure time, t^s, under constant load[177]

$$t_c = g^{-1} \cdot t_s \tag{34}$$

where t_s is the time to failure at a static stress equal to the average cyclic stress, and g^{-1} is a proportionality factor that depends on the type of stress cycle, its amplitude, and n. The factor g^{-1} can be calculated by numerical integration for any periodic cycle. For sinusoidal, square wave, and saw-tooth type loading, values of g^{-1} have been evaluated and are available in diagrams.[177] It has been verified in measurements that Equation 34 results in realistic lifetime predictions.[177,178]

If only linear cycles, i.e., saw tooth similar cycles, are considered, the lifetimes can be derived from a simple extension of the constant stress condition and the loading and unloading case with constant stress rate (Section IV.C.1 and 2).[158,175] This treatment holds for periodic as well as random cycling and leads to a lifetime

$$t_c = t_s / (n + 1) \tag{35}$$

where t_s means the failure time for a constant stress equal to the maximum stress of the cycles and n is the material constant. This relation can also be used to estimate the lifetime for nonlinear cycles; it describes the upper limit, whereas t_s determines the lower limit.[158,175] Such an estimation, however, will mostly be only rough because n is usually in the range of 50.

More complicated time-dependent load distributions must be treated by numerical methods. This procedure entails greater expenditure than any analytical solution, of course, but can be kept within reasonable limits if considering periodic distributions.[152] Such lifetime calculations have been performed, e.g., for different walking conditions.[157,189]

D. Probability of Failure — Lifetime Distributions

The equations for calculating lifetimes outlined in Section IV.C can be evaluated only if the initial flaw size is known. This knowledge, however, normally cannot be obtained by currently available nondestructive test methods.

On the other side, at least the distribution of the initial flaw sizes in a lot of ceramic specimens or components can be determined experimentally[176] by simple strength tests. According to Equation 27 the initial flaw size, a_i, is directly related to the inert strength, σ_{in}. If, therefore, the distribution of inert strengths of a lot of ceramic specimens is measured, and if it is assumed that this distribution can be expressed according to Weibull[147] as (Section III.D)

$$\ln \ln \frac{1}{1 - F} = m \cdot \ln \frac{\sigma_{in}}{\sigma_o} \tag{36}$$

then by replacing σ_{in} by Equation 27, a distribution of initial flaws can be obtained:

$$\ln \ln \frac{1}{1 - F} = - \frac{m}{2} \cdot \ln \frac{a_i}{a_{io}} \tag{37}$$

with $a_{io} = (K_{Ic}/f \cdot \sigma_o)^2$. If one now combines this equation with the corresponding equations for the failure times (Equations 24, 31, or 34), the lifetimes as functions of failure probability can be obtained.[175]

E. Fracture Parameter Measurements
1. Direct Methods
a. Crack Propagation Parameters

In order to measure the crack velocity, v, in dependence on the stress intensity factor, K, and to derive from it the crack propagation parameters, A and n (compare Equation 21), the double-torsion test turned out to be the most advantageous technique for ceramic materials.[152,157,170,182] It uses, on the one hand, a simple and material-saving specimen and allows, on the other hand, the crack length to be measured indirectly from the displacement of the specimen at the loading points. Details of this technique were discussed recently.[183,184]

b. Fracture Toughness

The most direct method to determine fracture toughnesses is that one according to the equation of definition Equation 16 for K_I. After introducing an artificial surface crack of defined size and shape — preferably after Equation 17 — the fracture stress is measured. This is usually performed in bending tests.

This procedure, however, often becomes difficult or even impossible in ceramics because the true crack length cannot be determined with sufficient accuracy. Then, specimens with different types of notches are used, either straight notches with varying radii where K_{Ic} follows from the extrapolation of the measured values to the radius zero[179] or with particularly shaped chevron notches.[180] For more details, one must be referred to the special literature where also the influences of the notch types and geometries on the resulting values are discussed.[180,181]

A further method, probably the best, but also the most expensive one, is to determine the fracture toughness from the V-K_I-curve. This curve becomes very steep for high velocities so that K_{Ic} can be evaluated with good accuracy (Compare Figure 4).

2. Indirect Methods
a. Crack Propagation Parameters

Under the assumption that the propagation of the microscopic cracks present in ceramic components can be described by the power function approach outlined in Section IV.B, the parameters A and n can be obtained also from static or dynamic strength measurements. This procedure could be advantageous because specimens are used which more closely resemble real structural parts than do specimens containing artificial cracks. This could mean that lifetime predictions based on static or dynamic fatigue data involve less extrapolation than do those based on crack velocity measurements.[175]

Under static fatigue conditions, the parameters are obtained by measuring the time to failure of a sufficiently large number of samples at several constantly applied stresses. From these data the median value of t_s can be calculated as a function of σ_a. After rewriting Equation 30 in the logarithmic form

$$\ln \bar{t}_s = \ln B + (n - 2) \cdot \ln \overline{\sigma}_{in} - n \cdot \ln \sigma_a \qquad (38)$$

n can be determined from a regression analysis of $\ln \bar{t}_s$ vs. $\ln \alpha_a$, B is determined from the intercept ($= \ln B + (n - 2 \cdot \ln \overline{\sigma}_{in})$) after measuring the median value of σ_{in} on another group of samples of the same lot.

Using dynamic fatigue, the parameters are obtained by a very similar procedure measuring fracture strength as a function of stressing rate $\dot{\sigma}$. After taking the logarithim in Equation 33 and rewriting it in terms of median values of σ_f and σ_{in} as

$$\ln \bar{\sigma}_f = \frac{1}{n+1} [\ln B + \ln(n+1) + (n-2) \ln \bar{\sigma}_{in} + \ln \dot{\sigma}] \tag{39}$$

n and B can be determined from a regression analysis of $\ln \bar{\sigma}_f$ vs. $\ln \dot{\sigma}$, the slope being equal to $1/(n+1)$, the intercept to $(\ln B + \ln (n+1) + (n-2) \cdot \ln \bar{\sigma}_{in})/(n+1)$.[175]

The median value technique for analyzing static and dynamic fatigue data is quite forward to apply. Other techniques which are regarded to make more efficient use of the data are discussed in detail in Reference 185. Uncertainties in lifetime predictions due to the experimental uncertainty in determining the fatigue parameters and the inert strength are discussed in References 16, 17, and 18.

b. Fracture Toughness

Knowing the crack propagation parameters by the indirect determinations described before, K_{Ic} can be estimated by extrapolating the power function which is actually valid only for the slow velocity range, up to velocities of some millimeters per second. This may not be correct, of course, because possible deviations of the curve are neglected, but it yields a conservative estimation and will be consistent for lifetime predictions based on the power function dependence alone.

3. Data

The range of the fracture mechanics parameters for the various ceramics is given in Tables 4 to 6. Evidently the fracture toughness of the different material groups shows the same tendency as the strength because both are correlated. The hydroxyapatites have the lowest crack resistance; most of the carbons are similar to them. Only some glass ceramics and pyrolytic carbons appear to attain higher values which could be compared with those of the weakest inert ceramics. The latter ones attain by far the highest fracture toughness; they normally differ by at least a factor of four from the other groups. For comparison it should be mentioned that the metallic materials such as steels or Co-based alloys range another order of magnitude higher.

In order to illustrate the meaning of the crack growth parameters given in Tables 4 to 6, some v-K_I-curves are depicted in Figures 5 and 6 which were obtained by direct measurement methods (dt). A log-log plot is chosen according to the relation

$$\log v = \log A + n \cdot \log K_I \tag{40}$$

which results from Equation 21 by taking the logarithm. In this representation n describes the slope of the curve, whereas log A determines its position in the diagram. It becomes evident that high n-values mean less time dependence of failure because only slight changes in K_I let the crack velocity increase drastically so that the critical crack length is reached very rapidly. In this way the high fatigue strengths which are observed for some carbon materials can be explained (compare Tables 1 and 4). On the other side, even n-values of about 50 which are typical for the inert ceramics, will lead to a significant time dependence.[96,97,157,189]

In Figure 5, the v-K_I-dependence of various materials which are taken as examples of the different groups is compared. They also show the differing resistance against crack propagation of the groups. In Figure 6, the influence of the physiological environment is demonstrated for different alumina bioceramics. As a main result it is to be seen that the crack

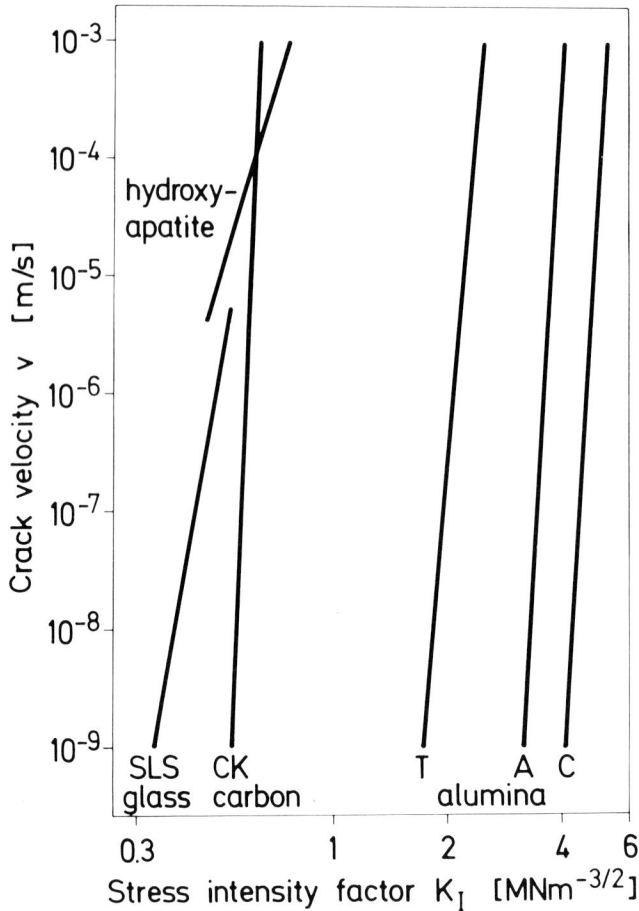

FIGURE 5.   v-K_I-dependences for various materials (hydroxyapatite,[155] soda-lime-silica-glass,[151] glassy carbon CK2,[152] technical dense alumina T,[152] alumina bioceramics A and C,[152] measured by double-torsion tests in air).

propagation is strongly enhanced by water or Ringer's solution. By this behavior the additional reduction of the long-term strength which has been found[99,159] becomes understandable as well as the reduced quasistatic strength values measured under these conditions.[145,159] For completion it must still be added that for all materials shown in Figures 5 and 6, except for glass, any deviation of the straight line, i.e., a Region II behavior (compare Figure 4), was not established in the crack velocity range investigated.

Further data are given in Tables 4 to 6. These data should show in more detail additional environmental and microstructural influences; however, the available knowledge seems to be poor. Out of the really biomedically used ceramics, only alumina is investigated more extensively, but also for this material the tendencies are not very clear; sometimes the results seem to be even contradictory. It appears to be obvious that aqueous solutions shift the v-K curve to smaller K_I-ranges.[152,155,157] Whether there is any difference between water and various physiological solutions which was found in one investigation[160] cannot be stated generally because other authors did not find any significant change.[152,159] Some increase in the crack resistance seems to exist if the density increases or if the porosity decreases, respectively[76,152] (compare also Figure 6 where the density increases from left to right). For different glassy carbons, however, an opposed dependence was found.[83] Furthermore, the

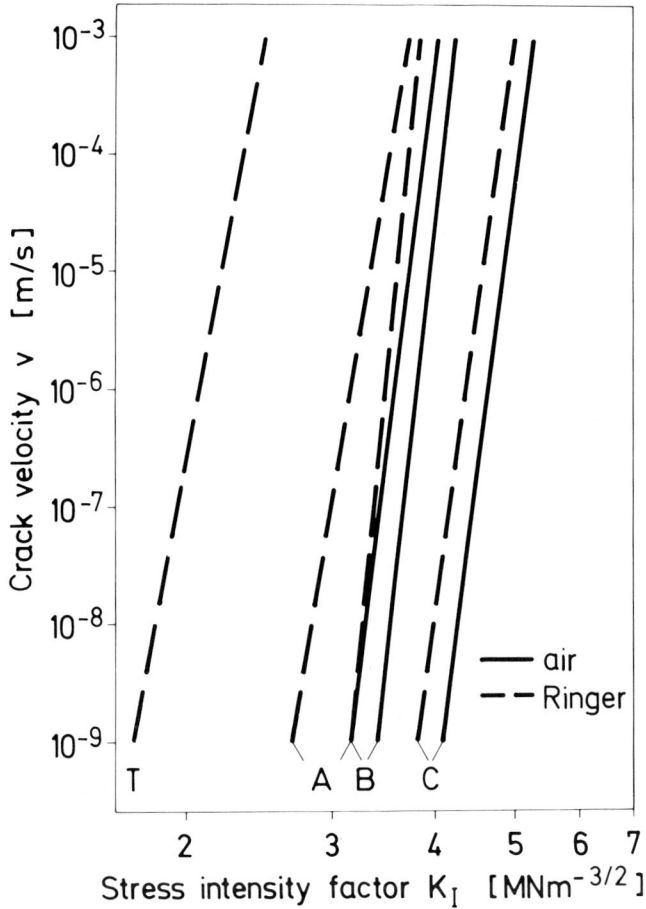

FIGURE 6.   Influence of simulated physiological environment on the v-K_I-dependence for different alumina.[142]

exponent, n, could be influenced by the grain size. At least in the case of alumina the materials with small grains (B and M8 in Table 6) show higher values than those with larger ones (A, C, and M9).[152,159] From these few findings it follows that additional investigations are necessary with respect to further possible improvements of the materials.

F. Proof Test

All the reliability or lifetime considerations described before are limited in a certain sense: either only statistical predictions with a remaining risk for failure are possible, or the initial flaw sizes must be known which appear to be very difficult or even impossible with regard to the techniques available today. These disadvantages can be overcome by applying a proof test.[158,173,174,190-193] The main idea of this procedure is to determine by a loading test of short duration the maximum effective flaw in a component and thus, to be able to calculate the minimum lifetime for given loading conditions in service.

In proof testing all components of a batch are subjected to a stress which is greater than that expected in service. The "proof stress", σ_p, must be chosen sufficiently high so that all components containing flaws $a > a_m$ which would lead to lifetimes shorter than desired will fail according to

$$K_{Ip} = \sigma_p \cdot \sqrt{a} \cdot f \geq K_{Ic} \qquad a \geq a_m \qquad (41)$$

and thus be eliminated. For all specimens that survive the test because of $a < a_m$ and therefore

$$K_{Ip} = \sigma_p \cdot \sqrt{a} \cdot f < K_{Ic} \qquad a < a_m \tag{42}$$

a minimum lifetime will be guaranteed.

The maximum allowable flaw size, a_m, follows from the different failure time equations derived in Section IV.C or from numerical calculations,[152] whereas the according proof stress level is defined by Equation 41. In the case of constant load in service, for instance, which simultaneously can be considered to estimate the worst case, the corresponding stress intensity factor K_{Ia} is

$$K_{Ia} = \sigma_a \cdot \sqrt{a} \cdot f$$

and thus, by combination with Equation 42

$$K_{Ia} < \frac{\sigma_a}{\sigma_p} K_{Ic} \tag{43}$$

Substituting K_{Ia} in Equation 26 the minimum time to failure in dependence on the proof stress can be calculated:[173,174]

$$t_{min} = \frac{2 \cdot K_{Ic}^{2-n}}{A(n-2) \cdot f^2 \cdot \sigma_a^2} \left(\frac{\sigma_p}{\sigma_a}\right)^{n-2} \tag{44}$$

or using Equation 29[158,175]

$$t_{min} = \frac{B}{\sigma_a^2} \left(\frac{\sigma_p}{\sigma_a}\right)^{n-2} \tag{45}$$

By means of such relations proof test diagrams can be plotted from which the minimum lifetime in dependence on the service load after application of a certain proof stress can be read.[158,173-175]

The proof test should be conducted under conditions where slow crack growth is minimized, i.e., in vacuum or in an inert environment. Crack growth during loading and at the proof stress does not affect the validity of Equation 44 or similar relations derived for different loading situations in service, whereas crack growth during unloading can invalidate the guarantee of performance. Therefore, rapid unloading rates must be used for which, however, the additional small increase in crack length can be estimated.[152] Further, the proof test must duplicate in the component the actual state of stress expected in service; and finally, after the proof test, the components must be protected from any subsequent mechanical damage.

It has been established by various investigations that the proof test concept principally works very well.[173,178,192,194] Sporadic deviations which have been sometimes observed, i.e., measured lifetimes shorter than calculated, can mostly be explained by inaccuracies in measurement or by the simplifying assumptions for the lifetime predictions.[193] Thus, proof testing is an excellent tool to improve and to assure the mechanical reliability of structural components.

V. CONCLUSIONS

In this chapter the mechanical properties and the strength behavior of the various ceramic

materials used for artificial biomedical devices have been discussed. These ceramics show usually lower strengths than other materials used in this field, Nevertheless, they are of extraordinary interest because this disadvantage is compensated by some other excellent properties with regard to the biological environment. Therefore, particular care must be taken in using them for structural components. This can be done only for uncritical applications by considering the conventionally determined strength properties. For higher safety requirements, improved methods of failure prediction should be used which are based, at least, on extended statistical evaluation or, as the best approach, on the fracture mechanics concept. In the latter case, combined with proof testing, a maximum mechanical reliability under given service conditions can be assured.

REFERENCES

1. **Hulbert, S. F., Klawitter, J. J., and Bowman, L. S.,** History of ceramic orthopaedic implants, *Mater. Res. Bull.,* 7, 1239, 1972.
2. **Heimke, G. and Griss, P.,** Ceramic implant materials, *Med. Biol. Eng. Comput.,* 18, 503, 1980.
3. **Heimke, G., Griss, P., Jentschura, G., and Werner, E.,** Bioinert and bioactive ceramics in orthopaedic surgery, in *Mechanical Properties of Biomaterials,* Hastings, G. W. and Williams, D. F., Eds., John Wiley & Sons, New York, 1980, 129.
4. **Bokros, J. C.,** Carbon in prosthetic devices, in *Trans. 4th Annu. Meet. Soc. for Biomaterials,* Vol. 2, Society for Biomaterials, San Antonio, Tex., 1978, 32.
5. **Hentrich, R. I., Graves, G. A., Stein, H. G., and Bajpai, P. K.,** An evaluation of inert and resorbable ceramics for future clinical orthopaedic applications, *J. Biomed. Mater. Res.,* 5, 25, 1971.
6. **Griss, P., von Andrian-Werburg, H., Krempien, B., and Hemike, G.,** Biological activity and histocompatibility of dense Al_2O_3/MgO ceramic implants in rats, *J. Biomed. Mater. Res. Symp.,* 4, 453, 1973.
7. **Griss, P., Krempien, B., von Andrian-Werburg, H., Heimke, G., and Fleiner, R.,** Experimentelle Untersuchung zur Gewebeverträglichkeit oxidkeramischer (Al_2O_3) Abriebteilchen, *Arch. Orthop. Unfall Chir.,* 76, 270, 1973.
8. **Griss, P., Werner, E., Budinger, R., Büsing, C. M., and Heimke, G.,** Zur Frage der unspezifischen Sarkomentstehung um Al_2O_3-keramische Implantate, *Arch. Orthop. Unfall Chir.,* 90, 29, 1977.
9. **Willert, H. G., Semlitsch, M., Buchhorn, G., and Kriete, U.,** Materialverschleiß und Gewebereaktion bei künstlichen Gelenken, *Orthopade,* 7, 62, 1978.
10. **Harms, J. and Mäusle, E.,** Tissue reaction to ceramic implant material, *J. Biomed. Mater. Res.,* 13, 67, 1979.
11. **Smith, L.,** Ceramic-plastic materials as a bone substitute, *Arch. Surg.,* 87, 653, 1963.
12. **Kenner, G. H., Pasco, W. D., Frakes, J. T., and Brown, S. D.,** Mechanical properties of calcia stabilized zirconia following in vivo and in vitro aging, *J. Biomed. Mater. Res. Symp.,* 6, 63, 1975.
13. **Hulbert, S. F., Young, F. A., Mathews, R. S., Klawitter, J. J., Talbert, C. D., and Stelling, F. H.,** Potential of ceramic materials as permanently implantable skeletal prostheses, *J. Biomed. Mater. Res.,* 4, 433, 1970.
14. **Hulbert, S. F., Morrison, S. J., and Klawitter, J. J.,** Tissue reaction to three ceramics of porous and non-porous structures, *J. Biomed. Mater. Res.,* 6, 347, 1972.
15. **Griss, P., Werner, E., and Heimke, G.,** Alumina ceramic, bioglass, and silicone nitride:a comparative biocompatibility study, in *Mechanical Properties of Biomaterials,* Hastings, G. W. and Williams, D. F., Eds., John Wiley & Sons, New York, 1980, 217.
16. **Dörre, E.,** Oxidkeramische Werkstoffe — ihre Eigenschaften und Anwendungen unter besonderer Berücksichtigung des Verschleißverhaltens, *VDI Ber. Nr.,* 194, 121, 1973.
17. **Dawihl, W., Mittelmeier, H., Dörre, E., Altmeyer, G., and Hanser, U.,** Zur Tribologie von Hüftgelenk-Endoprothesen aus Aluminiumoxidkeramik, *Med. Orthop. Tech.,* 99, 114, 1979.
18. **Heimke, G., Beisler, W., von Andrian-Werburg, H., Griss, P., and Krempien, B.,** Untersuchungen an Implantaten aus Al_2O_3-Keramik, *Ber. Dtsch. Keram. Ges.,* 50, 4, 1973.
19. **Ungethüm, M.,** *Technologische und biomechanische Aspekte der Hüft und Kniealloarthroplastik,* Huber, Bern, 1978, chap. 2.
20. **Dumbleton, J. H.,** *Tribology of Natural and Artificial Joints,* Elsevier, Amsterdam, 1981, chap. 7.
21. **Reckziegel, A. and Heimke, G., Oxidkeramik,** in Handb. der Keramik, *II K 1, Schmid, Freiburg i. Br.,* 1979.

22. **Maier, H. R., Pohlmann, H. J., and Krauth, A.,** Design criteria and structural testing of ceramic components, in *The Mechanical Engineering Properties and Applications of Ceramics,* No. 26, Godfrey, D. J., Ed., Proc. British Ceramic Society, Stroke-on-Trent, 1978, 17.

23. **Boutin, P.,** Arthroplastie totale de la hanche par prothèse en alumine frittée, *Rev. Chir. Orthop.,* 58, 229, 1972.

24. **Griss, P., Heimke, G., and von Andrian-Werburg, H.,** Die Aluminiumoxidkeramik-Metall-Verbundprothese. Eine neue Hüftgelenktotalendoprothese zur teilweise zementfreien Implantation, *Arch. Orthop. Unfall Chir.,* 81, 259, 1975.

25. **Salzer, M., Zweymüller, K., Locke, H., Plenk, H., Jr., and Punzet, G.,** Biokeramische Endoprothesen, *Med. Orthop. Tech.,* 95, 40, 1975.

26. **Semlitsch, M., Lehmann, M., Weber, H., Dörre, E., and Willert, H. G.,** Neue Perspektiven zu verlängerter Funktionsdauer künstlicher Hüftgelenke durch Werkstoffkombination Polyäthylen-Aluminiumoxidkeramik-Metall, *Med. Orthop. Tech.,* 96, 152, 1976.

27. **Geduldig, D., Lade, R., Prüssner, P., Willert, H. G., and Zichner, L.,** Tierexperimentelle Untersuchungen mit Gelenkendoprothesen aus dichter Al$_x$O$_3$-Keramik, *Med. Orthop. Tech.,* 96, 112, 1976.

28. **Schulte, W. and Heimke, G.,** Das Tübinger Sofortimplantat, *Die Quintessenz,* 26(6), 17, 1975.

29. **Mutschelknauss, E. and Dörre, E.,** Extensions-Implantate aus Aluminiumoxidkeramik, *Die Quintessenz,* 28(7), 1, 1977.

30. **Frenkel, G., Nowak, K., Schulz-Freywald, G., Bertram, K. J., Gruh, W., and Dörre, E.,** Untersuchungen mit nichtmetallischen Werkstoffen in der Zahn-, Mund- und Kieferchirurgie, *Dtsch. Zahnaerztl. Z.,* 32, 295, 1977.

31. **Hench, L. L. and Paschall, H. A.,** Direct chemical bond of bioactive glass-ceramic materials to bone and muscle, *J. Biomed. Mater. Res. Symp.,* 4, 25, 1973.

32. **Hench, L. L. and Paschall, H. A.,** Histochemical responses at a biomaterial's interface, *J. Biomed. Mater. Res. Symp.,* 5, 49, 1974.

33. **Blencke, B. A., Brömer, H., and Deutscher, K.,** Glaskeramik — ein neuer, bioaktiver Implantatwerkstoff, *Med. Orthop. Tech.,* 95, 144, 1975.

34. **Blencke, B. A., Brömer, H., Deutscher, K., and Pfeil, E.,** Glaskeramiken für Osteoplastik und Osteosynthese, Forschungsbericht BMFT-FB T 77-91, 1977.

35. **Eulenherger, J. and Niederer, P. G.,** Correlation of histology and interfacial shear strength of bioactive and bioinert implant materials, *Trans. 4th Annu. Meeting Soc. for Biomaterials,* Vol. 2, Society for Biomaterials, San Antonio, Tex., 1978, 141.

36. **Osborn, J. F. and Newesely, H.,** The material science of calcium phosphate ceramics, *Biomaterials,* 1, 108, 1980.

37. **Hench, L. L., Splinter, R. J., Allen, W. C., and Greenlee, T. K.,** Bonding mechanisms at the interface of ceramic prosthetic materials, *J. Biomed. Mater. Res. Symp.,* 2, 117, 1971.

38. **Strunz, V., Bunte, M., Stellmach, R., Gross, U. M., Kühl, K., Brömer, H., and Deutscher, K.,** Bioaktive Glaskeramik als Implantatmaterial in der Kieferchirurgie, *Dtsch. Zahnaerztl. Z.,* 32, 287, 1977.

39. **Bunte, M., Strunz, V., Gross, U. M., Brömer, H., and Deutscher, K.,** Vergleichende Untersuchungen über die Haftung verschiedener Materialien im Knochen, *Dtsch. Zahnaerztl. Z.,* 32, 825, 1977.

40. **Riedmüller, J. and Soltész, U.,** Modelluntersuchungen zur Spannungsverteilung in der Umgebung von Zahnimplantaten, *Zahnaerztl. Welt Zahnaeretl. Reform.,* 86, 842, 1977.

41. **Soltész, U., Siegele, D., and Riedmüller, J.,** Die Spannungsverteilungen um ein stufenförmiges Implantt im Modellversuch und im Vergleich zu einfachen Grundformen, *Dtsch. Zahnaerztl. Z.,* 36, 571, 1981.

42. **Hoffmann, F., Harnisch, J. P., Strunz, V., Bunte, M., Gross, U. M., Männer, K., Brömer, H., and Deutscher, K.,** Osteo-Keramo-Keratoproprothese: eine Modifikation der Osteo-Odonto-Keratoprothes nach Strampelli, *Klin. Monatsbl. Augenheilk.,* 173, 747, 1978.

43. **Greenspan, L. L. and Hench, L. L.,** Chemical and mechanical behavior of bioglass-coated alumina, *J. Biomed. Mater. Res. Symp.,* 7, 503, 1976.

44. **Griss, P., Greenspan, D. C., Heimke, G., Krempien, B., Buchinger, R., Hench, L. L., and Jentschura, G.,** Evaluation of a bioglass-coated Al$_2$O$_3$ total hip prosthesis in sheep, *J. Biomed. Mater. Res. Symp.,* 7, 511, 1976.

45. **Griss, P., Werner, E., Heimke, G., and Raute-Kreinsen, U.,** Vergleichende experimentelle Untersuchungen an Al$_2$O$_3$-Keramik und mit mod. Bioglas (L.L. Hench) beschichteter Al$_2$O$_3$-Keramik, *Arch. Orthop. Traumat. Surg.,* 92, 199, 1978.

46. **Strunz, V., Bunte, M., Gross, U. M., Männer, K., Brömer, H., and Deutscher, K.,** Beschichtung von Metallimplantaten mit bioaktiver Glaskeramik Ceravital, *Dtsch. Zahnaerztl. Z.,* 33, 862, 1978.

47. **Fuchs, G., Brömer, H., and Deutscher, K.,** Investigations with loaded glass ceramic-coated implants, Paper 3.7.5, presented at 1st World Biomaterials Congr., Baden b. Wien, April 8, 1980.

48. **Bhaskar, S. N., Brady, J. M., Getter, L., Grower, M. F., and Driskell, T. D.,** Biodegradable ceramic implants in bone, *Oral. Surg.,* 32, 336, 1971.

49. **Driskell, T. D., O'Hara, M. J., Sheets, H, D., Jr., Greene, G. W., Jr., Natiella, J. R., and Armitage, J.,** Development of ceramic and ceramic composite devices for maxillofacial applications, *J. Biomed. Mater. Res. Symp.,* 2, 345, 1972.

50. **Heide, H., Köster, K., and Lukas, H.,** Neuere Werkstoffe in der medizinischen Technik, *Chem. Ing. Tech.* 47, 327, 1975.

51. **Köster, K., Karbe, E., Kramer, A., Heide, H., and König, R.,** Experimenteller Knochenersatz durch resorbierbare Calciumphosphatkeramik, *Langenbecks Arch. Chir.,* 341, 77, 1976.

52. **Osborn, J. F. and Weiss, T.,** Hydroxylapatitkeramik - ein knochenähnlicher Biowerkstoff, *Schweiz. Monatsschr. Zahnheilkd.,* 88, 1166, 1978.

53. **Vermeiden, J. P. W., Rejda, B. B., Peelen, J. G. J., and de Grott, K.,** Histological evaluation of calcium hydroxyapatite bioceramics, pure and rein-forced with polyhydroxyethylmethacrylate, in *Evaluation of Biomaterials,* Winter, G. D., Leray, J. L., and de Groot, K., Eds., John Wiley & Sons, New York, 1980, 405.

54. **Cameron, H. U., MacNab, I., and Pilliar, R. M.,** Evaluation of a biodegradable ceramic, *J. Biomed. Mater. Res.,* 11, 179, 1977.

55. **Denissen, H. W., Rejda, B. V., and de Groot, K.,** Calciumhydroxyapatite/*p*-hydroxy-ethyl methacrylate (HA/p HEMA) composites as natural tooth-root-substitutes, *Trans. 4th Annu. Meeting Soc. for Biomaterials,* Vol. 2, Society for Biomaterials, San Antonio, Tex., 1978, 188.

56. **Riess, G.,** Klinische Erfahrungen mit Tricalciumphosphat-(TCP-) Implantaten, *Die Quintessenz,* 29, 19, 1978.

57. **Röhrle, H., Scholten, R., and Sollbach, W.,** Der Kraftfluβ bei neuartigen Hüftendoprothesen, in *Pauwels Symposium — Biomechanik in Orthopädie und Traumatologie,* Kölbel, R., Bergmann, G., and Rohlmann, A., Eds., Zentrale Universitätsdruckerei, Berlin, 1979, 165.

58. **Siegele, D. and Soltész, U.,** Influence of material stiffness on the stress distributions surrounding dental implants — a simplified finite-element study, paper presented at the Symposium on Head, Neck and Dental Implants, Amsterdam, October 2, 1981.

59. **Eder, F. X.,** *Moderne Meβmethoden der Physik,* Part 1, VEB Deutscher Verl. d. Wissenschaften, Berlin, 1968.

60. **Singer, E.,** Bestimmung der mechanischen und thermischen Eigenschaften von Elektrokeramik, in *Handb. der Keramik,* IV B 1 fb, Schmid, Freiburg i. Br., 1974.

61. *Standard Test Methods for Flexural Properties of Ceramic Whiteware Materials,* ASTM C 674, American Society for Testing and Materials, Philadelphia, 1977.

62. *Implants for surgery — Ceramic Materials Based on Alumina,* ISO 6474, International Organization of Standardization, Geneva, 1981.

63. *Chirurgische Implantate — Keramische Werkstoffe — Aluminiumoxidkeramik,* DIN 58 835, Deutsches Institut für Normung, Berlin, 1979.

64. **Roark, R. J. and Young, W. C.,** *Formulas for Stress and Strain,* 5th ed., McGraw-Hill, New York, 1975, chap. 7.

65. **Ogorkiewicz, R. M. and Mucci, P. E. R.,** Testing of fibre-plastics composites in three-point bending, *Composites,* 2, 139, 1971.

66. **Schreiber, E., Anderson, O. L., and Soga, N.,** *Elastic Constants and Their Measurement,* McGraw-Hill, New York, 1973.

67. *Standard Test Method for Young's Modulus, Shear Modulus, and Poisson's Ratio for Ceramic Whitewares by Resonance,* ANSI/ASTM C 848, American Society for Testing and Materials, Philadelphia, 1978.

68. *Annual Book of ASTM Standards,* part 17, American Society for Testing and Materials, Philadelphia, 1979.

69. **Mott, B. W.,** *Micro-Indentation Hardness Testing,* Butterworths, London, 1956.

70. **Page, T. F., Sawyer, G. R., Adewoye, O. O., and Wert, J. J.,** Hardness and wear behavior of SiC and Si_3N_4 ceramics, in *The Mechanical Engineering Properties and Applications of Ceramics,* No. 26, Goodfrey, D. J., Ed., Proc. British Ceramic Society, Stoke-on-Trent, 1978, 193.

71. **Sargent, P. M. and Page, T. F.,** The influence of microstructure on the microhardness of ceramic materials, in *The Mechanical Engineering Properties and Applications of Ceramics,* No. 26, Goodfrey, D. J., Ed., Proc. British Ceramic Society, Stoke-on-Trent, 1978, 209.

72. **Rice, R.W.,** Correlation of hardness with mechanical effects in ceramics, in *The Science of Hardness Testing and its Research Applications,* Westbrook, J. H. and Conrad, H., Eds., American Society for Metals, Metals Park, Ohio, 1973, 117.

73. **Brückmann, H.,** Kohlenstoff-Werkstoff in der Endoprothetik, *Biotechnische Umschau,* 2, 256, 1978.

74. Data sheets, Werkstoffe für Apparate und Anlagen, No. 12.76.2500 and 8.76.3000, Sigri Elektrographit GmbH, D 8901 Meitingen.

75. **Hempel, M.,** Dauerschwingverhalten von Kunstkohlewerkstoffen, *Arch. Eisenhuettenwes.,* 38, 55, 1967.

76. **Hodkinson, P. H. and Nadeau, J. S.,** Slow crack growth in graphite, *J. Mater. Sci.,* 10, 846, 1975.

77. **Olcott, E. L.,** Pyrolytic biocarbon materials, *J. Biomed. Mater. Res. Symp.,* 5, 209, 1974.

78. **Nilles, J. L. and Lapitsky, M.,** Biomechanical investigation of bone-porous carbon and porous metal interfaces, *J. Biomed. Mater. Res. Symp.,* 4, 63, 1973.

79. **Kaae, J. L.,** The mechanical properties of glassy and isotropic pyrolytic carbons, *J. Biomed. Mater. Res.,* 6, 279, 1972.

80. **Kaae, J. L.,** Structure and mechanical properties of isotropic pyrolytic carbons deposited below 1600°C, *J. Nucl. Mater.,* 38, 42, 1971.

81. **Kaae, J. L.,** The effect of annealing on the microstructures and the mechanical properties of poorly crystalline isotropic pyrolytic carbons, *Carbon,* 10, 691, 1972.

82. **Bokros, J. C.,** Carbon biomedical devices, *Carbon,* 15, 355, 1977.

83. **Minnear, W. P., Hollenbeck, T. M., Bradt, R. C., and Walker, P. L., Jr.,** Subcritical crack growth of glassy carbon in water, *J. Non Cryst. Solids,* 21, 107, 1976.

84. **Nadeau, J. S.,** Subcritical crack growth in vitreous carbon at room temperature, *J. Am. Ceram. Soc.,* 57, 303, 1974.

85. **Newesely, H.,** Implantatmaterialien, in *Zahnärztliche Werkstoffe und ihre Verarbeigung,* Eichner, K., Ed., Hütnig, Heidelberg, 1981, 249.

86. **Bullock, R. E. and Kaae, J. L.,** Size effect on the strength of glassy carbon, *J. Mater. Sci.,* 14, 920, 1979.

87. **Hucke, E. E., Fuys, R. A., and Craig, R. G.,** Glassy carbon: a potential dental implant material, *J. Biomed. Mater. Res. Symp.,* 4, 263, 1973.

88. **Piotrowsky, G., Hench, L. L., Allen, W. C., and Miller, G. J.,** Mechanical studies of the bone bioglass interfacial bond, *J. Biomed. Mater. Res. Symp.,* 6, 47, 1975.

89. **Pernot, F., Zarzycki, J., Bonnel, F., Rabischong, P., and Baldet, P.,** New glass-ceramic materials for prosthetic applications, *J. Mater. Sci.,* 14, 1694, 1979.

90. **Denissen, H. W. and de Groot, K.,** The response of the apatite ceramic surface to a simulated physiological environment, in *Dental Implants: Materials and Systems,* Heimke, G., Ed., Hanser, Munich, 1980, 35.

91. **Swart, J. G. N. and de Groot, K.,** Clinical experiences with sintered calciumphosphate as oral implant material, in *Dental Implants: Materials and Systems,* Heimke, G., Ed., Hanser, Munich, 1980, 97.

92. **Köster, K., Heide, H., and König, R.,** Resorbierbare Calciumphosphatekeramik im Tierexperiment unter Belastung, *Langenbecks Arch. Chir.,* 343, 173, 1977.

93. **Newesely, H.,** Metallimplantate oder keramische Implantate? — Eine werkstoffkundliche Gegenüberstellung, *Öst. Ztsch. Stomatol.,* 76, 204, 1979.

94. **Heimke, G., Griss, P., Frhr. von Andrian-Werburg, H., and Krempien, B.,** Aluminiumoxidkeramik, ein neues Biomaterial, *Arch. Orthop. Unfall Chir.,* 78, 216, 1974.

95. **Dörre, E.,** Aluminiumoxiderkeramik als Implantatwerkstoff, *Med. Orthop. Tech.,* 96, 104, 1976.

96. **Dawihl, W., Altmeyer, G., and Dörre, E.,** Statische und dynamische Dauerfestigkeit von Aluminiumoxid-Sinterkörpern, *Z. Werkstofftech.,* 8, 328, 1977.

97. **Maier, H. R., Stärk, N., and Krauth, A.,** Reliability of ceramic-metalic hip joints based on strength analysis, proof and structural testing, in *Mechanical Properties of Biomaterials,* Hastings, G. W. and Williams, D. F., Eds., John Wiley & Sons, New York, 1980, 177.

98. **Held, K. and Reckziegel, A.,** Oxidkeramik, in *Ullmanns Encyklop. der techn. Chemie,* Vol. 17, Verl. Chemie, Weinheim, 1979, 515.

99. **Frakes, J. T., Brown, S. D., and Kenner, G. H.,** Delayed failure and aging of porous alumina in water and physiological media, *Am. Ceram. Soc. Bull.,* 53, 183, 1974.

100. **Schüller, K. H.,** Magnesiumoxid, in *Handb. der Keramik,* II J 4c, Schmid, Freiburg i. Br., 1973.

101. **Salmang, H. and Scholze, H.,** *Die Physikalischen und Chemischen Grundlagen der Keramik,* Springer-Verlag, Berlin, 1968.

102. **Pampuch, R.,** *Ceramic Materials,* Elsevier, Amsterdam, 1976.

103. **Rice, R. W.,** Machining of ceramics, in *Ceramics for High-Performance Applications,* Burke, J. J., Gorum, A. E., and Katz, R. N., Eds., Brook Hill, Chestnut Hill, Mass., 1974, 287.

104. **Melko-Gabler, K.,** Zirkonoxid, in *Handb. der Keramik,* II J 4d, Schmid, Freiburg i.Br., 1973.

105. **Gugel, E.,** Nichtoxidkeramik, in *Handb. der Keramik,* II K2, Schmid, Freiburg i.Br., 1975.

106. Data sheet, Silicon carbide, infiltrated-material properties, B8 0978/1, Rosenthal Technik AG, D-8672 Selb.

107. Data sheet, Ceranox CS, D-3.1.80 FR, Annawerk, D-8633 Rödental.

108. Data sheet, Formkörper aus Siliziumkarbid, Sigri Elektrographit GmbH, D-8901 Meitingen.

109. Data sheet, Silicon nitride, hot pressed-material properties, B4 0978/le, Rosenthal Technik AG, D-8672 Selb.

110. Data sheet, Ceranox NH, D-3. 1.80 FR, Annawerk, D-8633 Rödental.

111. Data sheet, Silicon nitride, reaction bonded-material properties, B3 0978/le, Rosenthal Technik AG, D-8672 Selb.

112. Data sheet, Ceranox NR, D-3.1.80 FR, Annawerk, D-8633 Rödental.

113. **Arrol, W. J.,** The SIALONs — properties and fabrication, in *Ceramics for High-Performance Applications,* Burke, J. J., Gorum, A. E., and Katz, R. N., Eds., Brook Hill, Chestnut Hill, Mass., 1974, 729.

114. **Mitomo, M., Tanaka, Ii., and Muramatsu, K.,** The strength of α-sialon ceramics, *J. Mater. Sci.,* 15, 2661, 1980.

115. **Clarke, I. C., Phillips, W., McKellop, H. A., Moreland, J., and Amstutz, H. C.,** Sialon ceramic — a candidate material for total joint replacements in *Mechanical Properties of Biomaterials,* Hastings, G. W. and Williams, D. F., Eds., John Wiley & Sons, New York, 1980, 155.

116. **Arias, A.,** Pressureless sintered SIALON with low amounts of sintering aid, *J. Mater. Sci.,* 14, 1353, 1979.

117. **Schnittgrund, G. D., Kenner, G. H., and Brown, S. D.,** In vivo and in vitro changes in strength of orthopedic calcium aluminates, *J. Biomed. Mater. Res. Symp.,* 4, 435, 1973.

118. **McGee, T. D. and Wood, J. L.,** Calcium-phosphate magnesium-aluminate osteo-ceramics, *J. Biomed. Mater. Res. Symp.,* 5, 137, 1974.

119. **Akao, M., Aoki, H., and Kato, K.,** Mechanical properties of sintered hydroxyapatite for prosthetic applications, *J. Mater. Sci.,* 16, 809, 1981.

120. **Edington, J. W., Rowcliffe, D. J., and Henshall, J. L.,** The mechanical properties of silicon nitride and silicon carbide. I. Materials and strength, *Powder Metall. Int.,* 7, 82, 1975.

121. **Edington, J. W., Rowcliffe, D. J., and Henshall, J. L.,** The mechanical properties of silicon nitride and silicone carbide. II. Engineering properties, *Powder Metall. Int.,* 7, 136, 1975.

122. **Spriggs, R. M.,** Expression for effect of porosity on elastic modulus of polycrystalline refractory materials, particularly aluminium oxide, *J. Am. Ceram. Soc.,* 44, 628, 1961.

123. **Knudsen, F. P.,** Effect of porosity on Young's modulus of alumina, *J. Am. Ceram. Soc.,* 45, 94, 1962.

124. **Dilger, H.,** Untersuchungen der mechanischen und elektrischen Eigenschaften von Al_2O_3 mit ZnO- und NiO-Zusätzen bei hohen Temperaturen, *Ber. Dtsch. Keram. Ges.,* 51, 93, 1974.

125. **Rice, R. W.,** Effects of inhomogeneous porosity on elastic properties of ceramics, *J. Am. Ceram. Soc.,* 58, 458, 1975.

126. **Rice, R. W. and Donahue, T. J.,** Effect of inhomogeneous porosity distribution on elastic moduli of ceramics, *J. Am. Ceram. Soc.,* 62, 306, 1979.

127. **Godfrey, D. J. and Pitman, K. C.,** Some mechanical properties of silicon nitride ceramics: strength, hardness, and environmental effects, in *Ceramics for High-Performance Applications,* Burke, J. J., Gorum, A. E., and Katz, R. N., Eds., Brook Hill, Chestnut Hill, Mass., 1974, 425.

128. *Standard Test Method for Compressive (Crushing) Strength of Fired Whiteware Materials,* ASTM C 773, American Society for Testing and Materials, Philadelphia, 1974.

129. *Standard Test Method for Compressive (Crushing) Strength of Graphite,* ASTM C 695, American Society for Testing and Materials, Philadelphia, 1975.

130. *Standard Test Methods of Tension Testing of Carbon and Graphite Mechanical Materials,* ASTM C 565, American Society for Testing and Materials, Philadelphia, 1978.

131. **Maier, H. R.,** Bruchwahrscheinlichkeit von polykristallinem Aluminiumoxid unter Berücksichtigung von Volumen-, Oberflächen- und Belastungseinflüssen, Diss. TU Munich, 1974.

132. **Maier, H. R. and Heckel, K.,** Bruchwahrscheinlichkeit von polykristallinen Aluminiumoxid unter statischer Biege- und Zugbeanspruchung bei 800°C, *Ber. Dtsch. Keram. Ges.,* 54, 370, 1977.

133. **Maier, H. R., Nink, H., and Krauth, A.,** Statistische Festigkeitseigenschaften, Krafteinleitung und Bauteilzuverlässigkeit am Beispiel von reaktionsgebundenem Siliciumnitrid, *Ber. Dtsch. Keram. Ges.,* 54, 413, 1977.

134. **Shim, H. S.,** The behavior of isotropic carbons under cyclic loading, *Biomater. Med. Devices Artif. Org.,* 2(1), 55, 1974.

135. **Shim, H. S. and Schoen, F. J.,** The wear resistance of pure and silicon-alloyed isotropic carbons, *Biomater. Med. Devices Artif. Org.,* 2(2), 103, 1974.

136. **Ryshkewitch, E.,** Rigidity modulus of some pure oxide bodies, *J. Am. Ceram. Soc.,* 34, 322, 1951.

137. **Ryshkewitch, E.,** Compression strength of porous sintered alumina and zirconia, *J. Am. Ceram. Soc.,* 36, 65, 1953.

138. **Duckworth, W.,** Discussion of Ryshkewitch paper, *J. Am. Ceram. Soc.,* 36, 68, 1953.

139. **Springs, R. M., Mitchell, J. B., and Vasilos, T.,** Mechanical properties of pure, dense alumina oxide as a function of temperature and grain size, *J. Am. Ceram. Soc.,* 47, 323, 1964.

140. **Passmore, E.M., Spriggs, R. M., and Vasilos, T.,** Strength-grain size-porosity relations in alumina, *J. Am. Ceram. Soc.,* 48, 1, 1965.

141. **Knudsen, F. P.,** Dependence of mechanical strength of brittle polycrystalline specimens on porosity and grain size, *J. Am. Ceram. Soc.,* 42, 376, 1959.

142. **Carniglia, S. C.,** Reexamination of experimental strength-vs-grain-size data for ceramics, *J. Am. Ceram. Soc.,* 55, 243, 1972.

143. **Kirchner, H. P.,** *Strengthening of Ceramics — Treatments, Tests, and Design Applications,* Marcel Dekker, New York, 1979.

144. **Claussen, N., Steeb, J., and Pabst, R. F.,** Effect of induced microcracking on the fracture toughness of ceramics, *Ceram. Bull.,* 56, 559, 1977.

145. **Sinharoy, S., Levenson, L. L., Ballard, W. V., and Delbert, E. D.,** Surface segration of calcium in dense alumina exposed to steam and steam-CO, *Ceram. Bull.,* 57, 231, 1978.

146. **Krainess, F. E. and Knapp, W. J.,** Strength of a dense alumina ceramic after aging, *J. Biomed. Mater. Res.,* 12, 241, 1978.

147. **Weibull, W. A.,** A statistical distribution function of wide applicability, *J. Appl. Mech.,* 18, 293, 1951.

148. **Batdorf, S. B.,** Fundamentals of the statistical theory of fracture, in *Fracture Mechanics of Ceramics,* Vol. 3, Bradt, R.C., Hasselmann, D. P. H., and Lange, F. F., Eds., Plenum Press, New York, 1978, 1.

149. **Evans, A. G.,** A general approach for the statistical analysis of fracture, in *Fracture Mechanics of Ceramics,* Vol. 3, Bradt, R. C., Hasselmann, D. P. H., and Lange, F. F., Eds., Plenum Press, New York, 1978, 31.

150. **McClintock, F. A.,** Statistics of brittle fracture, in *Fracture Mechanics of Ceramics,* Vol. 1, Bradt, R. C., Hasselmann, D. P. H., and Lange, F. F., Eds., Plenum Press, New York, 1974, 93.

151. **Kerkof, F., Richter, H., and Stahn, D.,** Festigkeit von Gls — Zur Abhängigkeit von Belastungsdauer und verlauf, *Glastech. Ber.,* 54, 265, 1981.

152. **Richter, H., Seidelmann, U., and Soltész, U.,** Rißausbreitung in keramischen Knochenersatzwerkstoffen unter Simulation physiologischer Bedingungen, BMF T-Forschungsbericht T 82-003, 1982.

153. **Ritter, J. E., Jr. and Cavanagh, M. S.,** Fatigue resistance of a Lithium Aluminosilicate glass-ceramic, *J. Am. Ceram. Soc.,* 59, 57, 1976.

154. **Hing, P. and McMillan, P. W.,** The strength and fracture properties of glass-ceramics, *J. Mater. Sci.,* 8, 1041, 1973.

155. **De With, G., van Dijk, H. J. A., Hattu, N., and Prijs, K.,** Preparation, microstructure and mechanical properties of dense polycrystalline hydroxyapatite, *J. Mater. Sci.,* 16, 1592, 1981.

156. **Davidge, R. W., Tappin, G., and McLaren, J. R.,** Strength parameters relevant to engineering applications for reaction bonded silicon nitride and REFEL silicon carbide, *Powder Metall. Int.,* 8, 110, 1976.

157. **Richter, H., Seidelmann, U., and Soltész, U.,** Slow crack growth and failure prediction for alumina in physiological media, in *Evaluation of Biomaterials,* Proc. 1st Eur. Conf. on Eval. Biomaterials, Straßurg, 1977, Winter, G. D., Leray, J. L., and K. de Groot, Eds., John Wiley & Sons, New York, 1980, 227.

158. **Ritter, J. E., Jr., Greenspan, D. C., Palmer, R. A., and Hench, L. L.,** Use of fracture mechanics theory in lifetime predictions for alumina and bioglass-coated alumina, *J. Biomed. Mater. Res.,* 13, 251, 1979.

159. **Rockar, E. M., and Pletka, B. J.,** Fracture mechanics of alumina in a simulated biological environment, in *Fracture Mechanics of Ceramics,* Vol. 4, Bradt, R. C., Hasselmann, D. P. H., and Lange, F. F., Eds., Plenum Press, New York, 1978, 725.

160. **Ferber, M. K. and Brown, S. D.,** Subcritical crack growth in dense alumina exposed to physiological media, *J. Am. Ceram. Soc.,* 63, 424, 1980.

161. **Dalgleish, B. J. and Rawlings, R. D.,** A comparison of the mechanical behavior of aluminas in air and simulated body environments, *J. Biomed. Mater. Res.,* 15, 527, 1981.

162. **Li, Li-Shing, and Pabst, R. F.,** Subcritical crack growth in partially stabilized zirconia (PSZ), *J. Mater. Sci.,* 15, 2861, 1980.

163. **Jakus, K., Coyne, D. C., and Ritter, E. J., Jr.,** Analysis of fatigue data for lifetime predictions for ceramic materials, *J. Mater. Sci.,* 13, 2071, 1078.

164. **Lewis, M. H., Bhatti, A. R., Lumby, R.J., and North, B.,** The microstructure of sintered Si-Al-O-N ceramics, *J. Mater. Sci.,* 15, 103, 1980.

165. **Lumby, R. J., North, B., and Taylor, A. J.,** Properties of sintered sialons and some applications in metal handling and cutting, in *Ceramics for High Performance Applications* Vol. 2, Burke, J. J., Lenoe, E. N., and Katz, R. N., Eds., Brook Hill, Chestnut Hill, Mass., 1978, 893.

166. **Irwin, G. R.,** Fracture, in *Handbuch der Physik,* Vol. 6, Flügge, S., Ed., Springer-Verlag, Berlin, 1958, 551.

167. **Paris, P. C. and Sih, G. C.,** Stress analysis of cracks, in *Fracture Toughness Testing and its Applications,* ASTM STP 381, American Society for Testing and Materials, Philadelphia, 1965, 30.

168. **Rooke, D. P. and Cartwright, D. F.,** *Compendium of Stress Intensity Factors,* Hillingdon Press, Uxbridge, 1976.

169. **Evans, A. G.,** A method for evaluation the time-dependent failure characteristics of brittle materials — and its application to polycristalline alumina, *J. Mater. Sci.,* 7, 1137, 1972.

170. **Wiederhorn, S. M.,** Influence of water vapor on crack propagation in soda-lime glass, *J. Am. Ceram. Soc.,* 50, 407, 1967.

171. **Richter, H.,** Experimentelle Untersuchungenn zur Rißausbreitung in Spiegelglas im Geschwindigkeits- bereich 10^{-3} bis $5 \cdot 10^{3}$ mm/s, Dissertation, University of Karlsruhe, Karlsruhe, West Germany, 1974.

172. **Brown, S. D.,** Subcritical crack growth: a treatment based upon multibarrier kinetics, and an electrical network analog, in *Environmental Degradation of Engineering Materials,* Louthan, R. and McNitt, R. P., Eds., Virginia Polytechnic Inst., Blacksburg, V., 1977, 141.

173. **Evans, A. G., and Wiederhorn, S. M.,** Proof testing of ceramic materials — an analytical basis for failure prediction, *Int. J. Fract.,* 10, 379, 1974.

174. **Wiederhorn, S. M.,** Reliability, life prediction, and proof testing, in *Ceramics for High Performance Applications,* Burke, J. J., Gorum, A. E., and Katz, R. N., Eds., Brook Hill, Chestnut Hill, Mass., 1974, 633.

175. **Ritter, J. E., Jr.,** Engineering design and fatigue failure of brittle materials, in *Fracture Mechanics of Ceramics,* Vol. 4, Bradt, R. C., Hasselmann, D. P. H., and Lange, F. F., Eds., Plenum Press, New York, 1978, 667.

176. **Evans, A. G. and Johnson, H.,** The fracture stress and its dependence on slow crack growth, *J. Mater. Sci.,* 10, 214, 1975.

177. **Evans, A. G. and Fuller, E. R.,** Crack propagation in ceramic materials under cyclic loading conditions, *Metall. Trans.,* 5, 27, 1974.

178. **Richter, H. and Soltész, U.,** Proof-test an 3-Punkt-Biegeproben aus Al_2O_3-Keramik, in *Vorträge der 12. Sitzung des Arbeitskreises Bruchvorgänge, Freiburg 1980,* DVM, Berlin, 1981, 105.

179. **Pabst, R. F.,** Determination of K_{Ic}-values with diamond-saw-cuts in ceramic materials, in *Fracture Mechanics of Ceramics,* Vol. 2, Bradt, R. C., Hasselmann, D. P. H., and Lange, F. F., Eds., Plenum Press, New York, 1974, 555.

180. **Bretfeld, H., Kleinlein, F. W., Munz, D., Pabst, R. F., and Richter, H.,** Ermittlung des Bruchwiderstands an Oxidkeramik und Hartmetallen mit verschiedenen Methoden, *Z. Werkstofftech.,* 12, 167, 1981.

181. **Freiman, S. W. and Fuller, E. R., Jr., Eds.,** *Fracture Mechanics for Ceramics, Rocks, and Concrete,* ASTM STP 745, American Society for Testing and Materials, Philadelphia, 1981.

182. **Wiederhorn, S. M.,** Subcritical crack growth in ceramics, in *Fracture Mechanics of Ceramics,* Vol. 2, Bradt, R. C., Hasselmann, D. P. H., and Lange, F. F., Eds., Plenum Press, New York, 1974, 613.

183. **Fuller, E. R., Jr.,** An evaluation of double-torsion testing — analysis, in *Fracture Mechanics Applied to Brittle Materials,* Freiman, S. W., Ed., ASTM STP 678, American Society for Testing and Materials, Philadelphia, 1979, 3.

184. **Pletka, B. J., Fuller, E. R., Jr., and Koepke, B. G.,** An evaluation of double-torsion testing — experimental, in *Fracture Mechanics Applied to Brittle Materials,* Freiman, S. W., Ed., ASTM STP 678, American Society for Testing and Materials, Philadelphis, 1979, 19.

185. **Jakus, K., Coyne, D. C., and Ritter, J. E.,** Analysis of fatigue data for lifetime predictions for ceramic materials, *J. Mater. Sci.,* 13, 2071, 1978.

186. **Jacobs, D. F. and Ritter, J. E.,** Uncertainty in minimum lifetime predictions, *J. AM. Ceram. Soc.,* 59, 481, 1976.

187. **Wiederhorn, S. M., Fuller, E. R., Mandel, J., and Evans, A. G.,** An error analysis of failure prediction techniques derived from fracture mechanics, *J. Am. Ceram. Soc.,* 59, 403, 1976.

188. **Ritter, J. E., Bandyopadhyay, N., and Jakus, K.,** Statistical reproducibility of the crack propagation parameter N in dynamic fatigue tests, *J. Am. Ceram. Soc.,* 62, 542, 1979.

189. **Seidelmann, U., Richter, H., and Soltész, U.,** On the structural safety of ceramic hip-joint heads, in *Advances in Biomaterials,* Vol. 3, Winter, G. D., Gibbons, D. F., and Plenk, H., Jr., Eds., John Wiley & Sons, New York, 1982, 213.

190. **Wiederhorn, S.M.,** Prevention of failure in glass by proof-testing, *J. Am. Ceram. Soc.,* 56, 227, 1973.

191. **Ritter, J. E. and Wulf, S. A.,** Evaluation of proof testing to assure against delayed failure, *Am. Ceram. Soc. Bull.,* 57, 186, 1978.

192. **Ritter, J. E., Jr., Oates, P. B., Fuller, E. R., Jr., and Wiederhorn, S.M.,** Proof testing of ceramics. I. Experiment, *J. Mater. Sci.,* 15, 2275, 1980.

193. **Fuller, E. R., Jr., Wiederhorn, S. M., Ritter, J. E., Jr., and Oates, P. B.,** Proof testing of ceramics. II. Theory, *J. Mater. Sci.,* 15, 2282, 1980.

194. **Xavier, C. and Hübner, H. W.,** Proof testing of alumina to assure against premature failure, in *Science of Ceramics,* Vol. 2, Extended Abstracts, Stenungsund, 1981, 117.

.

Chapter 3

SHAPE MEMORY ALLOYS

R. Kousbroek

TABLE OF CONTENTS

I. INTRODUCTION

In the past few years the shape memory alloys have attracted the attention of metallurgists and design engineers because of a number of remarkable properties which open a revolutionary way of designing on the basis of entirely new principles compared with conventional alloys. The most striking features of this family of alloys are the shape memory effect, the pseudo-elasticity, and the very high damping capacity. A short definition of these effects will be given here for convenience, but a more extensive description of the shape memory effect and the pseudo-elasticity, which are important in the context of this chapter, will follow later.

Shape memory effect — This is the phenomenon by which, after an apparent plastic deformation, a metal alloy upon heating starts to remember its original shape at a certain temperature and returns to its deformed shape upon cooling.

Pseudo-elasticity — The effect by which a material recovers the induced "plastic" strain upon unloading is known as the pseudo-elasticity. The amount of this reversible strain is much greater than the classical, elastic strain. In constrast to the shape memory effect, the temperature remains constant.

High damping capacity — As a result of a change in internal structure, the damping capacity of these alloys can be varied considerably as a function of temperature, resulting in a very high damping capacity in certain temperature ranges.

These three features are associated with a martensitic transformation and this association limits the family of alloys exhibiting these special effects. In principle, all the alloys which exhibit a martensitic transformation are potential shape memory alloys, but experience shows that the effects only appear significantly in alloys having a reversible martensitic transformation, e.g., nickel- and copper-based alloys (e.g., Ti-Ni and Cu-Zn-Al).

Although the first basic information concerning the shape memory alloys was already observed about 40 years ago, it was in the 1960s that the usefulness was recognized. Since that time many potential applications of the shape memory alloys have been suggested.

In this review the potential uses of the shape memory alloys in medical applications will be highlighted (Section V). However, first, in order to understand fully the working principles of the biomedical devices, the basic metallurgical mechanisms of the shape memory alloys and the related effects will be discussed (Sections III and IV).

II. HISTORICAL BACKGROUND

The first observed shape memory phenomenon is the pseudo-elasticity. In 1932 Ölander observed this in a Au-Cd alloy and called it "rubber-like" behavior.[1] In the 1950s this phenomenon was also recognized in other alloys, e.g., In-Tl, Cu-Zn, and Cu-Al-Ni. Because of the great amount of reversible strain, this effect is also called "superelasticity". The maximum amount of reversible strain has been observed in a Cu-Al-Ni single crystal with a recoverable elastic strain of 24%.[2]

The first steps on the discovery of the shape memory effect were made in 1938 by Greninger and Mooradian,[3] observing the formation and disappearance of martensite with falling and rising temperature in a Cu-Zn alloy. However, this basic phenomenon of the memory effects, the thermoelastic behavior of the martensite phase, was first extensively studied 10 years later by Kurdjumov and Khandros.[4]

Since the first observation of the shape memory effect in Au-Cd in 1951,[5] the effect has also been reported in other alloys, e.g., Cu-Zn, In-Tl, Cu-Al-Ni, Ag-Cd, Ag-Zn, Cu-Al, Fe-Pt, Nb-Ti, and Ni-Al, but the great breakthrough came in the early 1960s, when Buehler et al.[6] of the U.S. Naval Ordnance Laboratory (now called the U.S. Naval Surface Weapons Center) discovered the shape memory effect in an equiatomic alloy of nickel and titanium, since then popularized under the name Nitinol (Nickel-Titanium Naval Ordnance Laboratory). With this alloy complete recovery of a maximum strain of 8% can be achieved by the shape memory effect, associated with a considerable force, which can perform work.[7,8] As Nitinol is difficult and expensive to manufacture and fabricate, the attention of metallurgists reverted to one of the first shape memory alloys Cu-Zn (brass). With the discovery that addition of small amounts of aluminium to brass raised its transition temperature considerably,[9] the shape memory effect of this new Cu-Zn-Al alloy could now be used in many practical applications at or near room temperature as was already possible with Nitinol. The great advantage of this Cu-Zn-Al alloy in comparison with Nitinol is that it is much cheaper and much easier to machine and fabricate. Since 1969, a major part of the fundamental research on Cu-Zn-Al shape memory alloys was done by Delaey et al.[10,11]

Several applications of the Ti-Ni and Cu-Zn-Al shape memory alloys have been developed, e.g., tube fitting systems, self-erectable structures, clamps, greenhouse window openers, thermostatic devices, different thermomechanical applications for automobiles, heat-engines, and biomedical applications. Several international symposia have been devoted exclusively to these alloys. In 1968 a first symposium concerning the shape memory alloy Nitinol was held in the U.S.[12] followed in 1975 by the first international symposium on shape memory effects in alloys and applications at Toronto, Canada.[13] Since 1976 shape memory alloys and the mechanisms are an ever recurring topic at such conferences as the International Conference on Martensitic Transformations (ICOMAT).

III. NATURE AND MECHANISM OF SHAPE MEMORY ALLOYS

The striking features of shape memory alloys are all closely related to the martensitic transformation. It is thus of value to describe the martensitic transformation and related phenomena first. It should also be noted that the exact nature and mechanism governing the behavior of shape memory alloys is not identical for all shape memory alloys.

A. Martensitic Transformation

Cohen et al.[14] formulated the next definition of martensitic transformation: ''A martensitic transformation is a lattice-distortive, virtually diffusionless structural change having a dominant deviatoric component and associated shape change such that strain energy dominates the kinetics and morphology during the transformation.'' If diffusion rules the transformation, atoms are changing places in the lattice in an uncoordinated way over long ranges. The resulting process can be fully described by the diffusion laws of Fick or by the method of Matano. However, if as is the case in the martensitic transformation, a diffusionless transformation occurs, a coordinated movement of large blocks of atoms is decisive for the resulting structure. The most well-known martensitic transformation is the phase transition responsible for the hardening of carbon steel caused by quenching after annealing at high temperature. In this case the austenitic, high-temperature fcc-structure changes into the martensitic, bct-structure on cooling.

From now on, regardless of the crystal structure, we will denote the high-temperature phase as the austenitic phase, while the product of the transformation will be called martensitic. In the present context, these terms are certainly not limited to ferrous materials.

Concomitant with the homogeneous lattice deformation, caused by the movement of the large blocks of atoms, dominant deviatoric shear displacements can cause an external measurable shape change. The associated strain energy will exert a dominant influence on the kinetics and morphology of the transformation. It is in this dominant influence of the strain energy on the growth characteristics that many nonferrous martensitic systems show the shape memory effect, whereas most ferrous martensites do not, because of differences in growth behavior. The effect of these differences in growth behavior on the potential of an alloy to exhibit the shape memory effect by transformation is related to the driving force at the onset of growth.[10] It is generally agreed that the reason for the transformation is the difference between the free energies of the austenitic and martensitic structure, i.e., one kind of interatomic bonding is energetically favored at lower temperatures, below some equilibrium temperature, T_o.

If the formation and disappearance of the martensitic phase is directly responsive to alternation in temperature, i.e., a thermoelastic behavior, the driving force is small (as in the nonferrous systems). while the stress necessary to induce growth is accordingly small. If the driving force at the onset of growth is large due to the absence of a thermoelastic equilibrium (the ferrous systems), growth of the martensite proceeds at high rates (of the order of the velocity of sound) which leads to a strongly disturbed, defective grain boundary which is subsequently essentially immobile. In this last case the shape memory effect appears for efficient application purposes between too broad and therefore unacceptable temperature boundaries (200 to 300°C).

B. Martensitic Morphology

Martensitic transformations can be induced by changes in temperature as well as by the application of stress. This can be explained by the following effects: (1) the free enthalpy of the austenitic and martensitic phase and so their equilibria depend not only on changes in temperature and composition, but also on stress; and (2) the nucleation and growth process are associated with shear strains and these will interact with stresses acting within, or applied to, the specimen. These thermodynamic and kinetic effects are strongly dependent on the direction of stresses with respect to the lattice orientations. Thus, two main groups of martensite can be recognized: (1) thermally induced martensite and (2) stress-induced martensite.

1. Thermally Induced Martensite

Thermally induced martensite is mainly characterized by its temperature dependence. It forms and grows continuously as the temperature is lowered and shrinks and vanishes continuously as the temperature is raised. In this case the transformation proceeds essentially in equilibrium between the chemical driving energy of the transformation and a resistive energy whose dominating component is the stored elastic energy. The transformation associated with this kind of thermally induced martensite is called a thermoelastic martensitic transformation and characterized by its transition temperature A_s, A_f, M_s and M_f (Figure 1A).

The growth rate appears to be governed solely by the rate of change in temperature.[15,16] However, if the transformation occurs spontaneously whenever the chemical driving energy largely exceeds the resistive energy (i.e., the growth rate is independent of the rate of temperature change), the resulting martensite is not called thermoelastic anymore, but burst martensite.

The thermoelastic, as well as the burst martensitic transformation, are frequently either partially of fully self-accommodating. The martensite forms either in zig-zag arrays (burst martensite), in packets (massive martensite), in groups, or in bands.

If the martensitic transformation is self-accommodating, the orientation of the growing martensite plate with respect to the orientation of its neighbor is the one that is energetically most stable in that particular strain field. The strain associated with one variant compensates the strain in the other variants. This requires the accommodation of all the local distortions involved in the formation of the individual lamellae with minimum macroscopic strain regardless of crystal direction. Because of the absence of sufficient time to permit any relaxation of the resulting local stresses, high densities of dislocations are observed in self-accommodating martensite. These dislocations are formed as the result of local strain energy

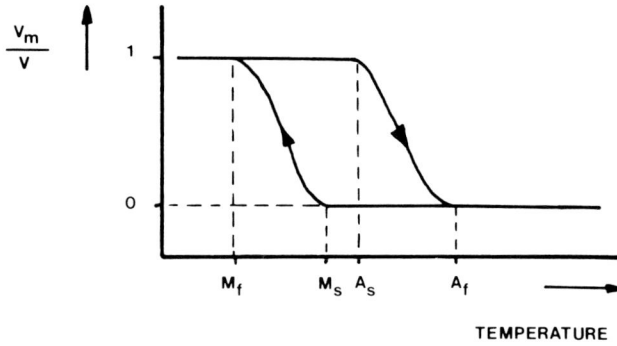

FIGURE 1 (A).   Thermoelastic martensitic transformation in function of temperature, with A_s = start transformation on heating; A_f = end transformation on heating; M_s = start transformation on cooling; M_f = end transformation on cooling, all at zero stress; and V_m/V = fractional volume of the martensite product.

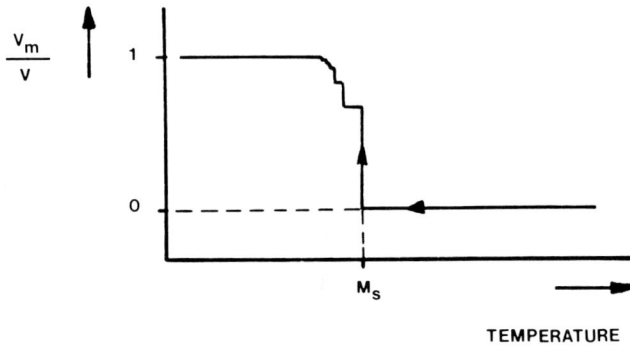

FIGURE 1 (B).   Schematical transformation of burst martensite on cooling. The initial bursts are expected to be large, decreasing rapidly as the martensitic fractional volume increases.[17]

as an alternative of the local breakdown of the lattice and not in response to any specific shear stress as is usually the case.

Therefore these dislocations are immobile and called accommodation dislocations, necessary to accommodate the microscopic transformation strain.[17] The maximum number of possible martensite orientations within one grain depends on the crystal symmetry of the austenitic phase. For a cubic austenitic phase, 24 martensite plate variants can occur.[15]

In the absence of external applied stresses and when the volume change is negligible, the thermally induced martensitic transformation is characterized by random martensite plate variants, resulting in a minimum or zero macroscopic shape deformation. However, if a constant external applied uniaxial stress assists the thermally induced martensitic transformation thermodynamically, only a limited number of thermoelastic martensite plate variants are expected to grow or, in the case of self-accommodating formations, certain martensite plate variants will become dominant in the different groups. This will lead to an external macroscopic shape change.

2. Stress-Induced Martensite

The stress-induced martensite is a mechanical analogue to the thermally induced martensite. In this case the transformation proceeds continuously with increasing applied uniaxial

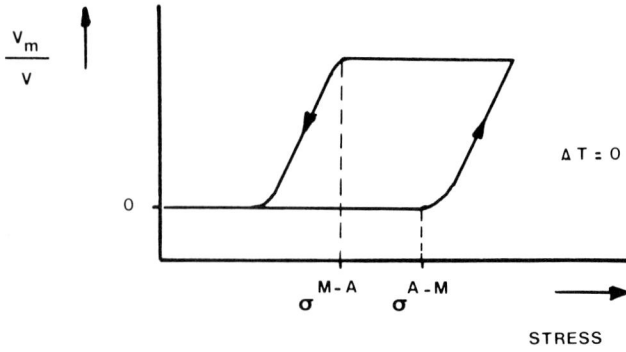

FIGURE 2.   Stress-induced, thermoelastic martensitic transformation in function of the applied stress, with $\sigma^{A \to M}$ = stress at which transformation austenite to martensite starts and $\sigma^{M \to A}$ = stress at which the reverse transformation martensite to austenite starts.

stress and is reversed continuously when the stress is decreased (Figure 2), while the temperature remains constant. Stress-induced martensite can be thermoelastic as well as burst-type martensite.

Again, those martensite plates will preferentially grow, which are most favorably oriented with regard to the externally applied uniaxial stress. Therefore, stress-induced martensite will have a strongly textured microstructure. The influence of external stresses on the martensitic transformation can be expressed by using the temperature M_s^σ, defined as the temperature at which the transformation to martensite can take place under an externally applied stress. The maximum temperature $M_s = (M_s^\sigma)_{max}$ depends on the stress conditions, specimen orientation or texture, the austenitic yield strength, and, possibly, other factors. It cannot be considered as an inherent characteristic of the material itself.[16]

If the deformation temperature, T_d, is approached on cooling and if $A_f < T_d < M_d$, martensite formed from the austenitic phase will disappear on the removal of the external stress. However, if $M_s < T_d < A_f$, the stress-induced martensite variants remain predominantly thermodynamically stable upon unloading. This includes that under stressed conditions the M_s-temperature is higher than in unstressed conditions.

3. Reoriented Martensite

If a fully thermally induced martensite structure is stressed at $T < M_f$ the martensite structure will be reoriented. This means that with regard to the applied stress certain preferential martensite variants will grow at the expense of the less suitably oriented martensite variants. The stress necessary to initiate the reorientation decreases with increasing number of transformation cycles, probably due to sweeping out existing defects along the martensite plate boundaries. Several theories are in circulation about the mechanism related to the reorientation. One proposed mechanism is the elastic twinning and untwinning in crystals, i.e., a reversible motion of existing twin boundaries resulting in a motion of the martensite plate boundary.[18] This implies that the thermally induced martensite has to be internally twinned.

Another mechanism for the reorientation of the martensite plates has been proposed by Wasilewski.[16,19,20] If the martensite is stressed at $T < M_f$ a stress-induced austenite variant, transformed out of the martensite, should be a transient intermediate transformation step, which is followed by the immediate and also stress-assisted transformation of this austenite to another martensite variant with another orientation with regard to the original martensite, i.e.,

$$M \overset{\sigma}{\rightarrow} \beta \overset{\sigma}{\rightarrow} M'$$

This transformation is limited between the upper temperature M_f and the lower temperature A_d, corresponding with the minimum temperature at which the transformation $M \overset{\sigma}{\rightleftharpoons} \beta$ can occur. The reverse transformation will occur in the opposite direction upon unloading. The existence of the intermediate step is a source of discussion. Other investigators describe a same kind of transformation by reorientation $M \rightarrow M'$ without this intermediate step.[21-23]

Although the exact theory is not yet clear, in crystallographic terms the two models are equivalent. Only the atomic position before and after the application of the stress can be considered, but not the path followed by the atoms.

Whether the reoriented martensite variant remains irreversible upon unloading or not, is dependent on the kind of shape memory alloy. If for a certain amount of strain the reoriented martensite is thermodynamically stable upon unloading, one has the situation required for one-way shape memory, induced by reorientation (Section III.D.1), while in the reverse the requirements are present for pseudo-elasticity by reorientation (Section III.E.2).

C. Crystallographic Requirements for Shape Memory Alloys

Originally the shape memory effect was ascribed to stress-assisted compositional changes,[6,24] but this theory did not stand firm for long. In 1966 the shape memory effect in Ti-Ni was for the first time related to a thermoelastic martensite transformation by Zijderveld et al.[25,26]

This association seemed to be too limited. Because in Ti-Ni, one has observed that 50 to 80% of the specimen apparently transforms instantaneously. This implies that the growth rate is not governed by the rate of change in temperature and therefore does not meet the definition of thermoelastic transformation, although Ti-Ni exhibits a very striking memory behavior.

A more general relation should be that any material exhibiting a forward and a reverse martensitic transformation, thermally or stress induced, is a potential shape memory alloy. It is only necessary to determine the exact conditions, associated with a complete and useful shape recovery, under which the martensitic transformation occurs.[17]

Another selection rule for shape memory alloys is the existence of an ordered structure.[18] It has been shown that the ordered Fe_3Pt-alloy exhibits a thermoelastic martensitic transformation and a related shape memory effect, while a disordered Fe_3Pt-alloy of the same composition displays neither.[18,27] This can be explained on the basis that the shape memory effect, as well as the pseudo-elasticity, originates from a complete crystallographic reversibility of the martensitic transformation. Otsuka and Shimizu[28] have shown that a path in the reverse transformation in ordered alloys is unique, in contrast to multiple paths in disordered alloys. They propose that the complete crystallographic reversibility of the martensitic transformation is characteristic of ordered alloys and rationalize the fact that the shape memory effect has usually been observed in ordered alloys. Besides, in the case of an ordering, irreversible plastic accommodation requires the creation of superdislocations, possessing energies which are multiples of those of the dislocations in the disordered lattice, while the matrix yield stress increases considerably as the result of the ordering (Figure 3).

Therefore, ordered alloys with their reversible martensitic transformation and their absence of lattice invariant plastic accommodation at the habit plane are much more favorable for shape memory purposes than the disordered alloys. There exists only one exception to this among shape memory alloys: the disordered In-Tl alloy.[18] However, here it involves a fcc $\leftrightarrows$ fct transformation, which is reversible even in disordered alloys. This stems from the fact that the lattice correspondence is unique in the reverse transformation because of the very simple lattice change and lower symmetry of the fct phase.[28]

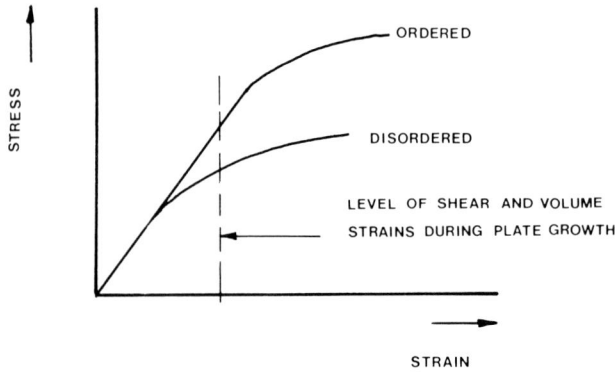

FIGURE 3.   Hypothetical stress-strain curves for ordered and disordered martensite. In the latter case transformation strains exceed the matrix (or martensite) elastic limit and interface coherence is lost.[18]

Another requirement for obtaining shape memory properties is that the martensite should be internally twinned.[18] If the deformed alloy has to revert to its initial state, e.g., when heated, the deformation process must be reversible, i.e., the shape memory alloy should not contain mobile dislocations. A possible mechanism for plastic deformation without mobile dislocations is internal detwinning of the martensitic substructure. The deformation can result from a selective detwinning process according to which one of the two twin orientations grows at the expense of the other. However, this theory seems subject to reservations. For example, Cu-Zn alloys are not internally twinned, yet they exhibit the shape memory effect.

Although not all the crystallographic requirements are conclusive, the shape memory effects are a reality. Fortunately, a detailed understanding of the physics of a process is not always necessary for successful exploitation.

D. Shape Memory Effect

When a conventional material is plastically deformed, a permanent deformation remains after unloading. The material can only regain its original shape by a new plastic deformation.

The shape memory alloys are distinguished from the "usual" alloys by the fact that, when these alloys are apparently plastically deformed, they remember their preceding shape on changing the temperature. Depending on the initial amount of deformation, they can regain their original shape (totally or partially) during heating. One can recognize two kinds of shape memory effects: (1) one-way shape memory effect and (2) two-way shape memory effect.

1. One-Way Shape Memory Effect

Figure 4 depicts schematically the one-way shape memory effect. The macroscopic deformation, which may not exceed a critical strain limit, is accompanied by a martensitic transformation, not reversed by removing the applied stress. If the specimen is fully martensitic at the onset of the deformation ($T_d < M_f$), the existing thermally induced martensite, M, is transformed by reorientation, according to one of the mechanisms of Section III.B.3, to the martensite variant M'. The reorientation takes place at the region BC of Figure 4. On unloading, the material recovers elastically, but a permanent deformation AD remains with a thermodynamically stable martensite structure. If the material is subsequently heated, the recovery of this deformation starts at the temperature A_s. The martensite variant obtained by reorientation reverts to the austenitic phase.

In the ideal case, strain recovery is complete if A_f is reached. For Nitinol, strains of the order of 6 to 8% may completely recover. If subsequent changes in temperature, in the

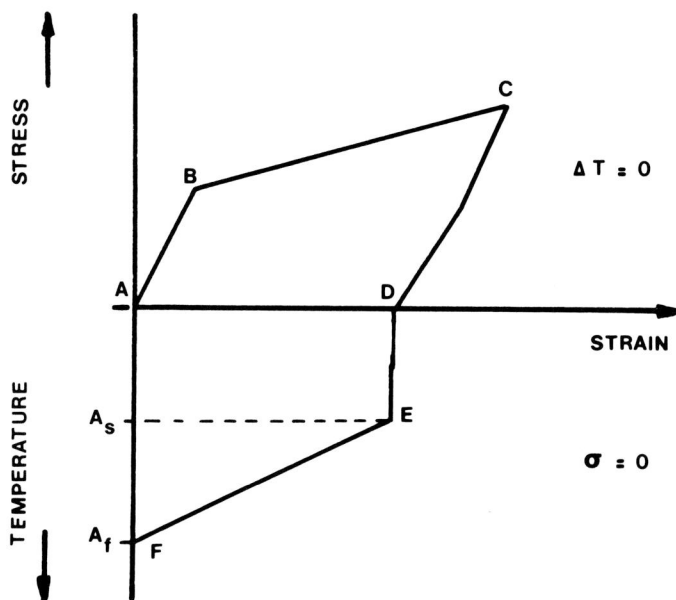

FIGURE 4. One-way shape memory effect.

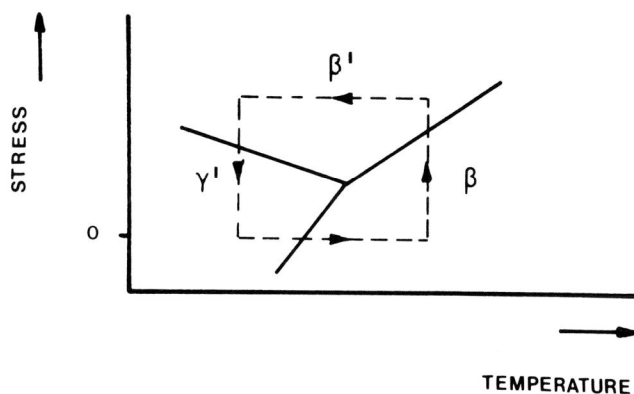

FIGURE 5. Stress-assisted transformation in Cu-Al-Ni.[29]

absence of external forces, do not change the macroscopic shape anymore, the phenomenon is called the one-way shape memory effect.

Another possibility is that the specimen is partially both martensitic and austenitic. If, for example, $M_f < T_d < M_s$ (T_d reached on cooling), the austenitic phase is transformed upon unloading to the irreversible stress-induced martensite variant, resulting in a predominantly martensitic structure, concomitant with a remaining deformation.

The last possiblity is that $T_d > A_f$, i.e., the specimen is fully austenitic. For Cu-Al-Ni the transformation path does not simply consist of a forward and a reverse transformation as usual, but consists of four steps[29] (Figure 5). By deformation of the austenitic structure β, stress-induced martensite β' will grow, resulting in a new γ'-martensite variant, con- comitant with a permanent strain upon cooling below M_f under constant stress (i.e., stress- assisted transformation) and after subsequent unloading. This strain can be annihilated by subsequent heating above the A_s-A_f range, where the γ'-martensite variant regains its original β-austenite structure.

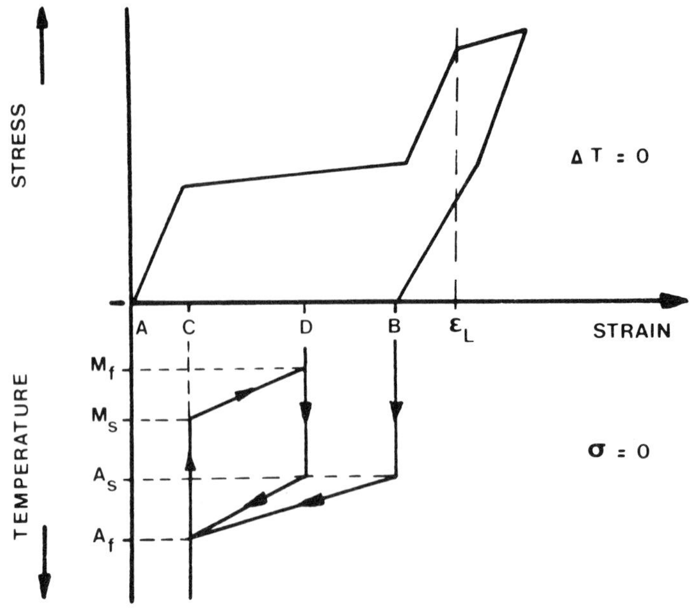

FIGURE 6.   Two-way shape memory effect.

A remarkable effect on heating is the appearance of an external measurable force, which can perform work. Because of the higher symmetry of the austenitic structure compared with that of the martensitic structure, the specimen strongly prefers to regain the austenitic state on heating. A part of the transformation energy is released as this recovery stress.

2. Two-Way Shape Memory Effect

In contrast with the one-way shape memory effect, subsequent cooling does indeed influence the macroscopic shape, while no external forces are applied. As well as the material remembering its undeformed shape on heating, it also remembers the deformed shape on cooling. Figure 6 depicts this phenomenon schematically. After deformation and unloading the material, a permanent deformation, AB, remains. Due to an initial nonuniform deformation (e.g., bending) above a critical strain ϵ_L (for Nitinol, $\epsilon_L = 8\%$), a plastic deformation of the martensitic structure, stress-induced or reoriented, is introduced, associated with internal stresses and preferential nucleation sites. These stresses and nucleation sites control the growth of a very select number of strongly textured martensite variants.

Due to the nonuniform deformation, some fibers in the material will exceed the critical deformation limit, while others do not. On heating, the less deformed fibers try to regain their original length, but are constrained by the highly deformed fibers. This results macroscopically in a reversible residual strain, AC, and microscopically in a residual stress pattern. On cooling below M_s this residual stress pattern will induce strongly textured martensite with appropriate orientation to accommodate the internal stresses. Macroscopically, the material seems to remember the deformed shape and will, again partially, return to it (point D). In subsequent temperature cycles, the shape of the material will vary between the low-temperature shape C and the high-temperature shape D, unless the temperature reached during the heating cycles is sufficient to relieve the residual stress, in which case the two-way shape memory effect will deteriorate.[30]

In addition to the introduction of the two-way memory effect by straining above a certain critical strain limit, it also can be obtained by training. Training is accomplished by limiting the number of martensite variants, formed when an alloy is repeatedly heated and cooled

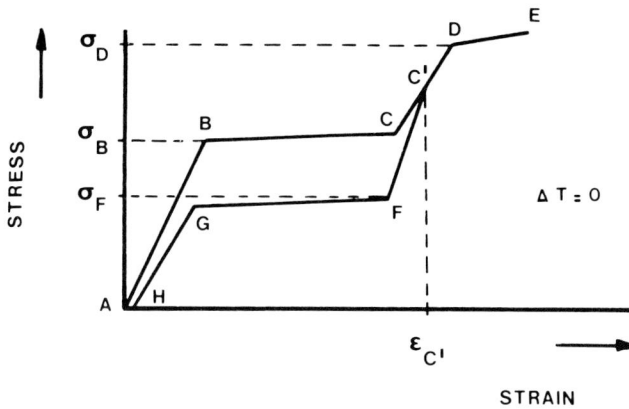

FIGURE 7. Pseudo-elasticity.

through the transition temperature range. At every cycle, the material is deformed in the low-temperature state, i.e., after a certain number of one-way shape memory effect cycles with intermediate deformation the effect tends towards two-way shape memory.[11,31]

Note that as described above the changes in external shape as a function of temperature occur within a selected, relatively narrow temperature range. This distinguishes the shape memory alloy in behavior from a bimetal, which also exhibits a visible shape change related to changes in temperature. However, this is a continuous change over all the imposed temperature range. Another difference between them is that the shape memory alloys can bridge a much greater change in shape compared with the bimetals within a selected temperature range, while the force exerted by shape memory alloys is also much greater.

E. Pseudo-Elasticity

Apparently, this phenomenon looks like an elastic deformation, and because of the non-elastic looking stress-strain curve the phenomenon has been called pseudo-elasticity (or superelasticity due to the large reversible strains). However, this phenomenon has nothing to do with a conventional elastic deformation, but exclusively with a stress-induced, thermoelastic transformation. The pseudo-elasticity is a complete mechanical analogue to the thermally induced shape memory effect. In this case the change in shape varies with varying applied stress. Strains up to 15% in polycrystalline material can be isothermally completely recovered by releasing the externally applied stress.[32] Figure 7 shows a schematic representation of this behavior in terms of a stress-strain curve. Three initial states of the material can be differentiated, each with its own pseudo-elastic mechanism.

1. Pseudo-Elasticity by Transformation

The initial structure consists of the austenitic phase. On stressing at a constant temperature ($A_f < T_d < M_d$), the austenitic phase transforms to a martensitic structure. In Figure 7, section AB represents purely elastic deformation of the austenitic phase. The stress, corresponding to point B, is the minimum stress at which the first stress-induced martensite plates start to form ($\sigma_B = \sigma_T^{A \rightarrow M}$). The transformation is completed when point C is reached. The difference between the slopes of sections AB and BC indicates the ease with which the transformation occurs. After the transformation is completed, the martensitic structure is deformed elastically on continued stressing, represented by section CB. At point D the yield stress of the martensitic phase is reached ($\sigma_D = \sigma_Y^M$) and the material deforms plastically until fracture occurs at E. However, if the stress is released before reaching point D, e.g., at point C', the strain $\epsilon_{C'}$ is recovered in several stages. First, elastic unloading of the

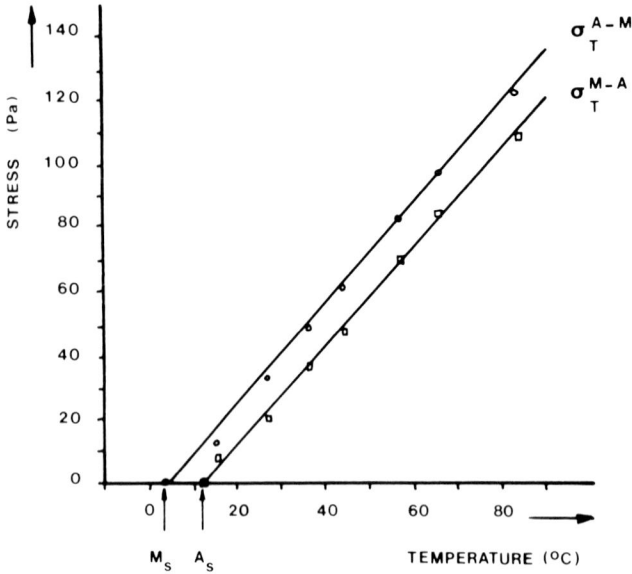

FIGURE 8.   Effect of temperature upon the tensile behavior of Cu-Zn-Si alloy.[32]

martensite occurs, represented by section C′F. At point F, the corresponding stress is the maximum stress on which the stress-induced martensitic structure on unloading can exist, so at this point the reverse transformation martensite to austenite starts ($\sigma_F = \sigma_T^{M \rightarrow A}$) and the fraction of martensite decreases until the austenitic structure is completely restored (point G). Section GH represents the elastic unloading of the austenitic phase. The total strain may or may not be completely recovered, depending on some irreversible deformation taking place either during loading or unloading.

The value of $\sigma_T^{A \rightarrow M}$ and $\sigma_T^{M \rightarrow A}$ is dependent on the deformation temperature and the crystal orientation of the austenitic structure, as well as whether the structure is polycrystalline or a single crystal. An example for these values of a polycrystalline Cu-Zn-Si alloy is given in Figure 8.[32] Although tensile or compressive stresses are mainly responsible for pseudo-elasticity, the macroscopic behavior is more pronounced on bending. This is caused by the fact that a small linear strain gives rise to a large bending strain in thin specimens.

2. Pseudo-Elasticity by Reorientation

The initial structure is martensitic. On stressing thermally induced martensite at a low temperature ($T_d < M_f$), pseudo-elasticity can also occur by reorientation of the martensite according to one of the mechanisms proposed in Section III.B.3. In constrast with what has been said in Section III.D.1, the reoriented martensite is now thermodynamically unstable on unloading and will revert to its original orientation after unloading.

After an elastic deformation of the initial existing martensite structure, the stress reaches the level at which the reorientation of the initial martensitic structure starts into a variant of more favorable orientation ($\sigma = \sigma^{M \rightarrow M'}$). After unloading, the reoriented martensite variant reverts to its original orientation at a certain stress-level ($\sigma = \sigma^{M' \rightarrow M}$) until the original variant is completely restored. The value of $\sigma^{M \rightarrow M'}$ and $\sigma^{M' \rightarrow M}$ decreases with the number of cycles.

3. Pseudo-Elasticity by Transformation and Reorientation

Finally, a third way to obtain pseudo-elasticity is the appearance of the unstable stress-induced martensite obtained by transformation at $T_d > A_f$. This is depicted in Figure 9.

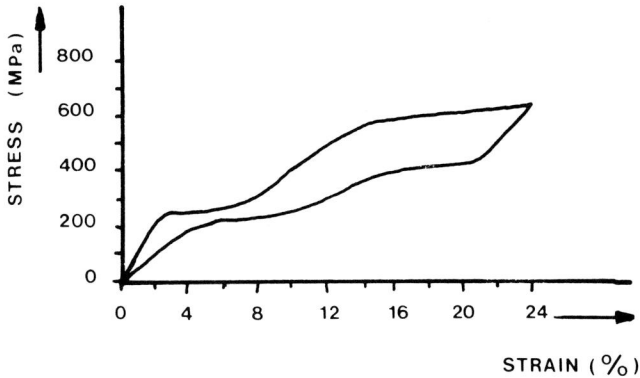

FIGURE 9.   Stress-strain curve for a single crystal of Cu-Al-Ni.[2]

First, the austenite transforms to the thermodynamically unstable stress-induced martensite variant. After this transformation is completed the martensite is stressed elastically until the martensite is reoriented. Upon unloading, both stages of the process are reversed. In this case a pseudo-elastic strain up to 24% can be achieved.

F. Overview of the Coherence between Structure and Memory Effects

Figures 10A and B give a schematic view of the coherence between the structure at the deformation temperature, T_d, and the mechanism responsible for inducing pseudo-elasticity and the shape memory effect in respect of the sequence of transformation temperatures.

A difference has been made in approaching the deformation temperature, T_d. Once again it has to be said that the appearance of a mechanism and its related memory effect depends strongly on the kind of memory alloy. Nevertheless, there exists a very close relationship between pseudo-elasticity and the shape memory effect because of the similarities in mechanisms. It has been proved that pseudo-elasticity and the shape memory effect are entirely complementary, i.e., if one effect is small, the other will be large and vice versa. The total recovery (pseudo-elastic plus shape memory) is close to 100% except for polycrystals above A_s.[33]

IV. STRUCTURE RELATED MECHANICAL PROPERTIES

A. Young's Modulus and Yield-Stress

As a function of the existing structure, the mechanical properties are totally different. For Nitinol, a transition in properties has been observed at the vicinity of the transformation temperature.[7]

Figure 11 indicates that the Young's modulus as well as the yield stress are much higher in value in the austentic state than in the martensitic state. The variation in the Young's modulus can be related to the critical chemical driving force around M_s.[34] The smaller the magnitude of this chemical driving force, the smaller will be the elastic strain energy that must be stored between the two existing phases to maintain thermoelastic balance. A low elastic strain energy involves a low Young's modulus. Since the chemical driving force increases on heating above M_s the Young's modulus will also increase drastically. Both Cross et al.[7] and Golestaneh[35] found for Nitinol a four times higher value for the Young's modulus in the austenitic structure than in the martensitic structure. A similar temperature dependence exists for the yield-stress due to changes in elastic constants in the vicinity of the transformation temperature.[36]

The change in properties implies that the material in the high-temperature, austenitic state is much stiffer, while at the same time, higher stresses can be accommodated than in the

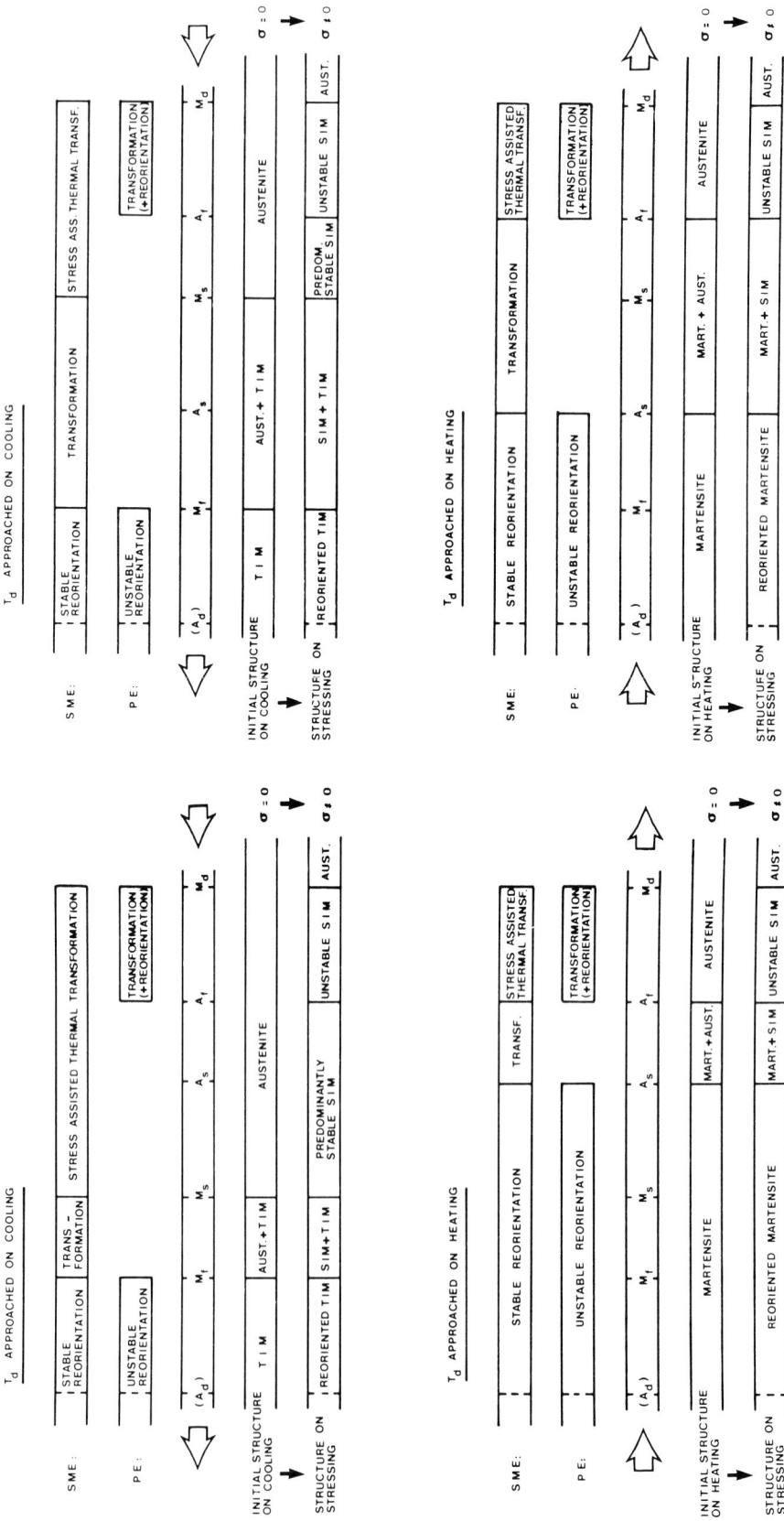

FIGURE 10 (A).   Structure-mechanism relationship if $M_s < A_s$.

FIGURE 10 (B).   Structure-mechanism relationship if $M_s > A_s$.

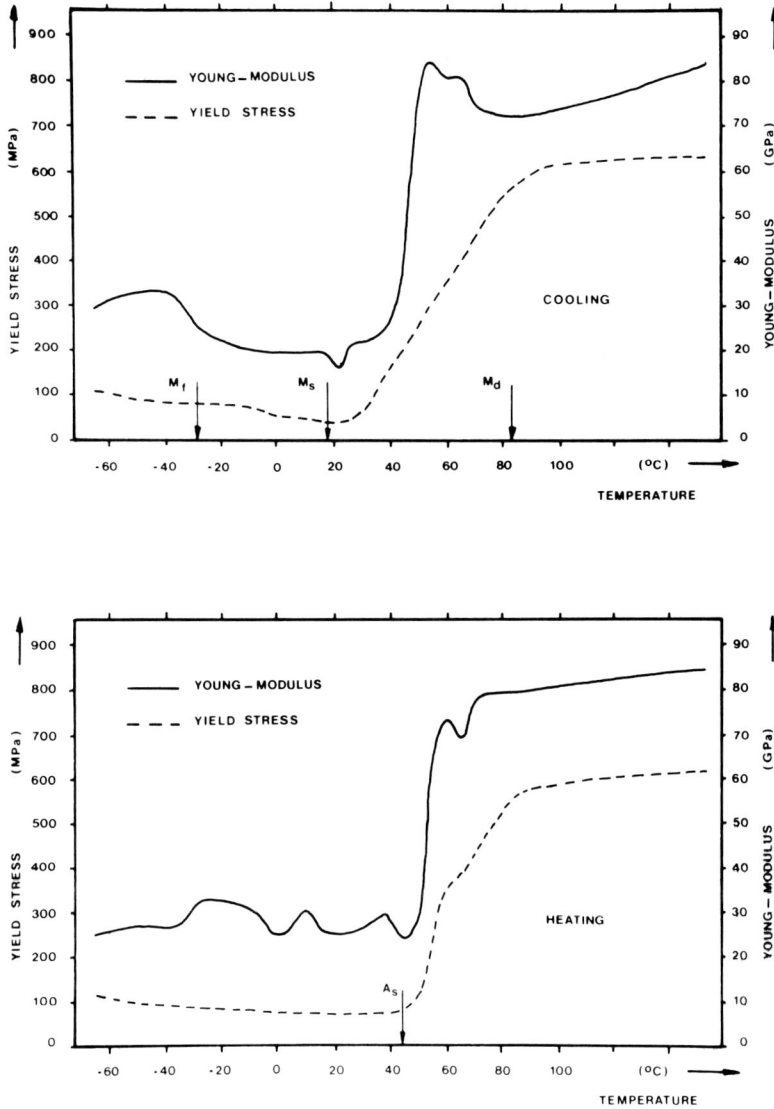

FIGURE 11. Young's modulus and yield-stress in function of temperature.[7]

low-temperature martensitic state. From the design point of view, this can be useful. If the material resists a high stress in its high-temperature state, on cooling below the M_s-temperature, the material loses its high stiffness and progressively deforms under the applied stress. This can be used, for example, to trigger temperature-dependent release mechanisms.

B. Recovery Stress

As reported earlier (Section III.D.1), an external measurable stress will be developed during heating, if the initial strain is restrained to prevent reversion, for example, by an external weight. This stress is called the recovery stress σ_r and will increase in sigmoidal fashion over the A_s-A_f range, as depicted in Figure 12 as a function of the initial strain. For Nitinol, the maximum recovery stress is reached at an initial strain of 8%.[7] Note that the A_s-A_f range is generally broadened and displaced to higher temperatures by the stress.

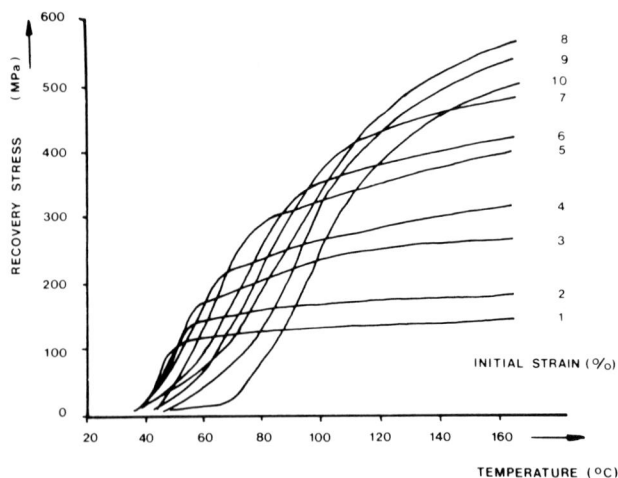

FIGURE 12.   Tensile recovery stress in function of temperature.[7]

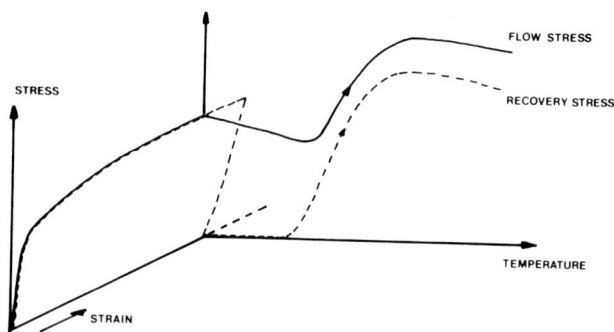

FIGURE 13.   Recovery-stress compared with temperature dependence of flow-stress.

The limit on attainable recovery stress at a certain temperature in a prestrained shape memory alloy is the flow stress of the austenitic phase at that temperature.[37] Figure 13 depicts the relationship between the recovery stress and the flow stress.

If the recovery stress equals the externally imposed stress, the development of the recovery stress will stop and, if the transformation is not completed, a displacement caused by the shape memory effect will occur, shown by lifting an external weight. If this happens, the shape memory alloy performs a certain amount of work which can be applied usefully.

C. Shape Memory Fatigue

Little has been published on the fatigue properties of shape memory alloys in respect to the great number of potential memory alloys. Publications are known on the fatigue properties of Au-Cd,[38] Cu-Zn-Sn,[39] Cu-Al-Ni,[40,41] Cu-Zn-Al,[42-44] and Nitinol.[7,20,45,46] Because Cu-Zn-Al and Nitinol seem to possess the most promising fatigue behavior, the properties of these two alloys have been compared.[47,48]

Most of these experiments concern the fatigue properties of the alloys in martensitic or austenitic state at a constant temperature. They describe the fatigue behavior of the pseudo-elasticity both by reorientation and by transformation if, respectively, $A_d < T_f < M_f$ and $A_f < T < M_d$. An important difference in fatigue behavior can be noticed as a function of:

1. Geometry of the specimen (round tensile specimens possess a better fatigue behavior than flat specimens)
2. Test method (three load modes are currently in use, the tension-compression cycling, the pulsating tension cycling, and the rotating bending fatigue test)
3. Grain size and texture
4. Structure remaining after each cycle

Of less importance is the absolute test temperature; of more importance is the relationship between test temperature and transformation temperature range.

The fatigue strength of Nitinol in the martensitic structure, reported as 480 MPa at a fatigue life greater than 10^7 c, accompanied by considerable specimen deflection during testing (standard R. R. Moore Test),[7] is anomalously high compared with its yield strength in the same structure ($\sim$100 MPa, see Figure 10).

A causal relationship between the martensitic transformation and the high fatigue strength has been proposed.[20] The apparently anomalous high fatigue strength of Nitinol should be due to the formation of very high accommodation dislocation densities in the early fatigue cycles. Because of the presence of these dislocations, little or no plastic deformation in the conventional sense, i.e., dislocation movement, is likely to be involved.

The Cu-Zn-Al alloys show better fatigue behavior in the martensite structure than in the austenitic phase.[42] Although in comparison with Nitinol the Cu-Zn-Al alloys possess worse fatigue properties at comparable stresses due to intergranular cracking resulting from strain incompatibilities across the grain boundaries, the resistance to high strain fatigue is superior to that of most other metals. The fatigue properties of Cu-Zn-Al are seriously improved by grain size refinement combined to a suitable texture[43] and alloying with additional elements.

Besides the fatigue properties at constant temperature, the fatigue limits at alternating temperatures through the transformation temperature range are important, i.e., the repeatability of the shape memory effect. This part of the research on fatigue properties of shape memory alloys seems to be neglected. Only in a few publications is some attention given to this phenomenon.[7,49,50] For a Cu-Zn-Ga alloy, it has been demonstrated that when the two-way shape memory effect is introduced by a training cycle, the magnitude of the two-way shape memory effect decreases relatively quickly during the initial 30 thermal c, but reaches an almost constant value after 50 to 100 c.[49]

An almost identical behavior has been demonstrated for Nitinol, under constant straining at every start of a new thermal cycle (i.e., repeated one-way shape memory effect). Depending on the initial degree of strain, stability will be reached between 40 to 80 c.[7] Nitinol wires in linear tension have been shown to withstand tens of thousands of cycles without permanent elongation.[50] However, further research in this domain will be necessary for obtaining reliable shape memory devices for long-term application.

V. SHAPE MEMORY ALLOYS FOR MEDICAL APPLICATION

Returning to the aim of this book, only Nitinol can act as a real biomaterial, e.g., a material that comes into contact with and is tolerated by the living tissue without any deterioration in properties and in its specific function after implantation within living systems or incorporation with them. However, besides internal applications, shape memory alloys can be used in external biomedical devices where biocompatibility requirements are less stringent. This indicates that because of their remarkable properties and the relative simplicity of inducing them, many interesting applications of the shape memory alloys are in the field of external devices. In the recent past, some applications in different disciplines of medicine have been proposed, all but one based on the shape memory effect. Because of their most efficient shape memory behavior, until now only Nitinol and Cu-Zn-Al are used in the biomedical shape memory devices.

A. Biocompatibility

Up to now, only Nitinol was shown to possess good biocompatibility behavior. It is therefore, preeminently suitable for internal biomedical application. To determine the biocompatibility of Nitinol for implant applications, both "in vivo" and "in vitro" experiments were carried out. The first reported biocompatibility test on Nitinol was done in 1973 by Cutright et al.,[51] followed by those of Castleman et al.,[52-54] Hughes,[55] Schmerling et al.,[56] and Haasters et al.[57]

Cutright et al.[51] examined Nitinol wire implanted subcutaneously in rats during a 9-week period. There appeared to be no difference in reaction compared with stainless steel similarly implanted for the same period.

Nitinol bone plates were implanted into the femurs of beagles by Castleman et al.,[52,53] using Nitinol screws and instruments for fastening them. In addition, a number of Vitallium® bone plates were implanted into femurs of other beagles as controls. The bone plates were removed from the animals and examined after implantation of 3, 6, 12, and 17 months. There were no signs of generalized or localized corrosion attributable to a reaction between the metallic surfaces and immediately adjacent tissue. There was no significant difference in the histological observations of bone and tissue adjacent to the implants, whether made of Nitinol or Vitallium®. Neither was there any histopathological deviation in the removed liver, spleen, lung, kidney, and brain.

Both Motzkin et al.[54] and Hughes[55] have carried out tissue culture studies on Nitinol, using human fibroblasts and buffered fetal rat calvaria tissue. Pure nickel and titanium, 316-L stainless steel, and Vitallium® have been used as controls. Both research groups found that the Nitinol results are well within the limits of acceptability. Additional "in vivo" experiments on mice and sheep,[55] rhesus monkeys,[56] as well as on rats,[57] all with no evidence of foreign body rejection or inflammation, confirmed the opinion that Nitinol is as biocompatible as presently accepted materials.

If the shape memory effect is applied "in vivo" in direct contact with the tissue, one has to ensure that the alloy exhibits a transformation temperature range in which no tissue damage can be caused. This temperature range is about 10 to 45°C.

B. Design Considerations

From the design point of view, the shape memory alloys offer attractive advantages in comparison with more conventional alloys. As was described earlier, the memory effects can be induced at constant temperature or at alternating temperatures through the transformation temperature range.

At constant temperature, pseudo-elasticity can be utilized in the design. Large reversible strains can be induced, while the stress to induce this strain hardly increases above the apparent yield stress (see Figure 7). If temperature changes are incorporated into design considerations, the shape memory effect can be applied. For internal biomedical engineering, the great advantage is that the devices can be implanted in an optimum shape for surgery, after which the desired functional shape is obtained *in situ* within the body.

The shape memory effect can be involved in different ways in the design, considering either the ability of the alloy to return to different shapes upon temperature cycling or the force that is generated and the mechanical work that can be done by the alloys as it attempts to regain its present shape upon heating. Depending on the initial structure two modes of use are possible.

1.   The material is deformed in the martensitic structure. On heating above the A_s-temperature, the original shape is recovered, or, if the device is constrained, a recovery stress will be generated.

A combination of these two features is also possible if the shape memory element is connected with a freely movable load. On the first change in temperature upon heating, the shape memory element can be considered as being constrained by the load. Subsequently, a recovery stress will be generated while no shape change can occur. If the recovery stress equals the external stress, imposed by the load, the increase of the recovery stress stops and a motion of the shape memory element starts. This motion can be both linear and torsional, resulting in lifting up the external load. In this way, the shape memory element can perform mechanical work.

2.   The material is in the austenitic structure and resists a high stress. On cooling below the M_s-temperature, the material becomes weaker by decreasing of the Young's modulus and will progressively deform under the applied stress.

If one intends to use the two-way shape memory element as a precision actuator, the hysteresis between change in shape upon heating and cooling can be a disadvantage. Much of this hysteresis can be suppressed by forcing the actuator to work against a bias spring.[58] With a bias spring the shape memory element can be "trimmed" to perform in that part of the load-deflection-temperature spectrum where significant amounts of energy can be harnessed to carry out the desired function. However, some restrictions have to be made if the (T_{min}-T_{max}) range is within the transformation temperature range. Because of the incomplete transformation the shape memory effect can decrease and at last vanishes totally.

In the case where input power for heating is limited, only small volumes of shape memory material may be used. Because of the correspondingly small amount of energy available, the shape memory effect may best be used to trigger the release of other stored, mechanical energy.[59]

C. Potential Applications

At the end of the 1960s the first potential applications of the Nitinol shape memory alloy in bioengineering were proposed.[60,61] These applications were spread over a wide spectrum of medical disciplines:

1.   Devices to restrain inaccessible body parts during surgery[60]
2.   A temperature recording indicator for use in blood shipments, possibly combined with a self-regulating value system to release refrigerant into a storage system, ensuring constant temperature in shipment[61]
3.   Small motor to power artificial heart[61]
4.   Device to ease intermittent occlusion[61]
5.   Construction of artificial muscles[61]
6.   Self-tightening wire for jaw fixation following fracture of mandible[61]
7.   Device to aid reduction and assure approximation of long bone fractures[61]
8.   Intrauterine contraceptive device[62]
9.   Kinetic memory electrodes, catheters, and cannulae[63]

Some of the applications proposed in recent years will be described more extensively, classified into four different disciplines of medicine.

1. Dentistry and Orthodontics

The only commercially available medical application of Nitinol is in this medical discipline. It concerns an orthodontic dental arch wire for straightening malpositioned teeth, marketed by Unitek Corporation under the name Nitinol Activ-Arch® wire.[64] This type of arch wire, which is attached to bands on the teeth, is intended to replace the traditional stainless steel arch wire. Although recently efforts have been made to use the shape memory

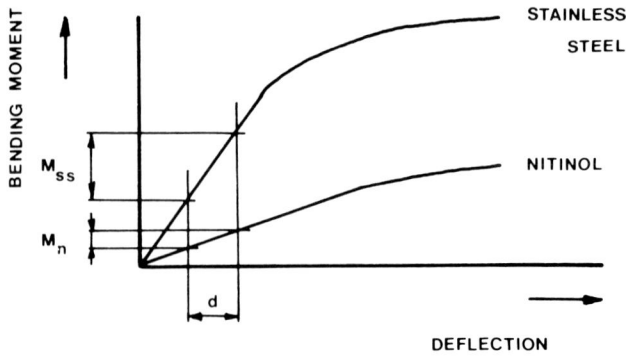

FIGURE 14.   Comparison between Nitinol and stainless steel wire.[66]

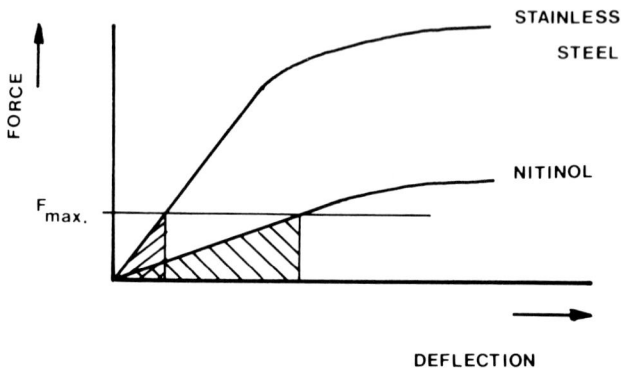

FIGURE 15.   Comparison stored energy.[61]

effect in orthodontic wires,[65] the working principle of this Nitinol Activ-Arch® wire is not the shape memory effect nor the pseudo-elasticity, but the relatively low Young's modulus of Nitinol in the martensitic condition (~30 GPa). This modulus is low in comparison with the modulus of stainless steel (~200 GPa).[66] In combination with a cold work process, when the Nitinol has been drawn into a high-strength wire, the alloy exhibits an outstanding elasticity. Figure 14 shows a comparison of the change in bending moment between Nitinol and stainless steel wire with the same dimensions when undergoing a constant change in deflection.

The stainless steel wire undergoes a much larger change in moment compared to the change in moment of the Nitinol wire, when both wires are being deflected an equivalent amount. Clinically, this means that for any given malocclusion Nitinol wire will produce a lower, more constant force on the teeth than would a stainless steel wire of equivalent size. On the other hand, if the force is chosen low enough so that both wires are capable of achieving full recovery without any plastic deformation, the stored energy of the Nitinol wire is much greater than the stored energy of a similar stainless steel wire (Figure 15).

It is this increased energy and so its greater working range than the stainless steel wire that accounts for increased clinical efficiency of Nitinol because of the fewer required arch wire adjustments in the course of treatment, reducing the chair time. The advantages for patients are the shorter treatment time required to accomplish rotations and leveling and less discomfort because of the applied low, continuous force instead of heavy, intermittent force if treated with stainless steel.

Another application in dentistry is the use of Nitinol as a jaw osteosynthetic material using the shape memory effect. Such devices were proposed by Civjan et al.[67] and Bensmann et al.[68] The latter developed a shape memory jaw plate, acting as a compression plate upon heating. The compression stress at the fracture site appears to be higher than in the conventional osteosynthetic plates. Dependent on the plate geometry, the ratio of increase is from 1.36 to 2.48.

2. Orthopedics

Dynamic compression bone plates exhibiting the shape memory effect are one of the most favorite orthopedic applications of Nitinol,[55,69,70] followed by the intramedullary fixation nail.[55,68,71,72] The fracture healing in long bones can be accelerated when bone ends are held in good position by means of compression between the bone fragments. A widely accepted method uses an internally applied dynamic compression bone plate. The plate is affixed by screws to one side of the fractured bone with the plate overlapping the fracture site and extending to the other side of the fractured bone. On screwing the other side down to the bone a relative motion between the screws and the plate in the axial direction will occur, pulling the fracture sites together, and thereby effecting compression and rigid fixation to the bone fragments.

With the shape memory effect in the Nitinol dynamic compression bone plate, the compression is effected by the return of the prestrained plate to its original shape. With this method undesirable surface damages and wear of the holes in the conventional dynamic bone plate is avoided, while a continuous compression is assured, even if bone resorption occurs at the fracture sites. This effect remains as long as the original shape is not reached.

Another possible application in fracture healing in long bones is intramedullary fixation. Instead of the previously mentioned cortical dynamic compression bone plate, in this case a nail is driven into the reamed medullary canal, stabilizing the bone fragments with regard to each other. However, this can raise some problems if the nail has to be removed after fracture healing or if bone resorption occurs in the medullary canal.

If the jamming of the nail in the medullary canal is too tight, it is possible that the nail cannot be removed after fracture healing without causing damage to the bone structure. With a nail exhibiting the two-way shape memory effect, this can be avoided. If the outer diameter of the predeformed nail is less than the diameter of the reamed medullary canal, insertion will be easy. Upon warming to body heat, the nail tries to regain its original greater diameter thus causing a tight jamming in the medullary canal. Upon cooling after fracture healing, the nail returns to its small diameter shape so that it can be easily extracted.

On the other hand, excessive initial jamming causes bone resorption in the medullary canal, resulting in reduction of fixation and increased rotation of the bone fragments, preventing healing. With a modified design of an intramedullary fixation nail with shape memory effect fixation elements, all these problems can be avoided, while an additional compression between the fractured bone ends can be effected. Figure 16 shows the principle of the modified intramedullary fixation nail, proposed by the present author,[71] consisting of independently movable shape memory effect fixation elements attached to two profiled polymer rods, interconnected by a shape memory effect compression rod.

In the design of this intramedullary fixation nail the following conditions were considered:

1. At low temperature ($T < M_f$), the nail should possess a suitable shape, making insertion and, if necessary, extraction of the nail easy.
2. At high temperature ($A_f < T \leq 37°C$), the nail should fit strongly in the fractured bone without causing damage to the bone structure.
3. In addition to a tight diametral fit of the nail at "high" temperature, the nail must also apply a certain amount of compression in order to press the fractured surfaces together.

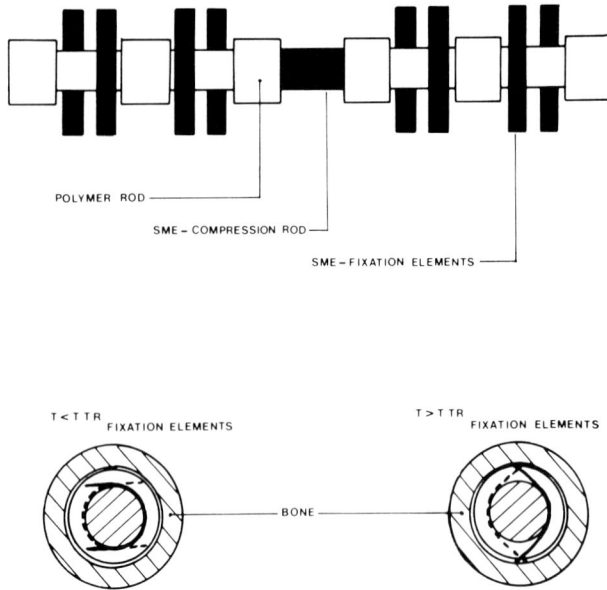

FIGURE 16.   Intramedullary fixation nail.

The independently movable fixation elements can follow the local, actual diameter of the medullary canal, ensuring a tight fit of the intermedullary nail, even in the case of an irregularly shaped medullary canal or after bone resorption, while the compression rod accelerates the fracture healing.

For optimal compression, the fixation elements and the compression rod should possess different transformation temperature range (TTR), e.g.,

$$(A_f)_{\text{FIXATION ELEMENTS}} \leq (A_s)_{\text{COMPRESSION ROD}}$$

and

$$(A_f)_{\text{COMPRESSION ROD}} = \text{max } 30°C$$

This ensures that axial compression will start only after the fixation elements are firmly fitted to the bone and will be accomplished before body temperature has been reached, as is depicted in Figure 17. This modified intramedullary fixation nail is only a preliminary design. The final design of this fixation nail will be published elsewhere.[72] Other potential applications in orthopedics are a total hip prosthesis fixation system,[55] a total surface replacement arthroplasty,[68] and devices for the treatment of scoliosis, using the Harrington method[56] or the method of Dwyer.[73]

3. Rehabilitation

In the field of external appliance, the shape memory effect and the pseudo-elasticity can be used to simplify dynamic rehabilitation devices as far as the moving parts are concerned. Besides Nitinol other shape memory alloys can be used because no biocompatibility is required. The transformation temperature range is less critical and mainly dependent on the applied heating and cooling system in comparison with internal applications.

A first application of the shape memory effect in rehabilitation engineering has been proposed by Kousbroek et al.[71] Here also it concerns a preliminary result, while the final design will be published elsewhere.[72]

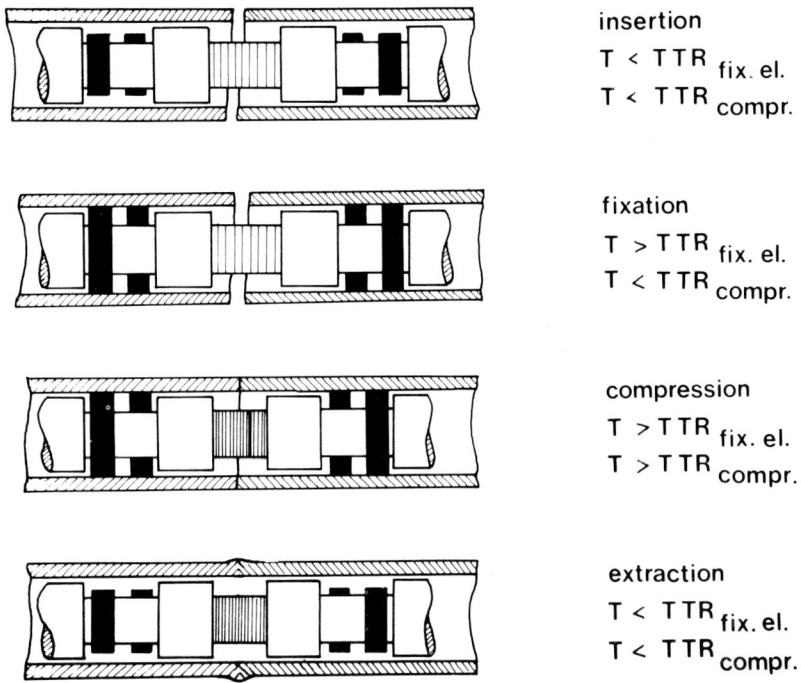

insertion

$T < TTR_{\text{fix. el.}}$

$T < TTR_{\text{compr.}}$

fixation

$T > TTR_{\text{fix. el.}}$

$T < TTR_{\text{compr.}}$

compression

$T > TTR_{\text{fix. el.}}$

$T > TTR_{\text{compr.}}$

extraction

$T < TTR_{\text{fix. el.}}$

$T < TTR_{\text{compr.}}$

FIGURE 17.   Sequence of transformation.

A dynamic handsplint has been designed in the shape of a glove on which a flexion or an extension motion can be imposed by means of fixed strips of Cu-Zn-Al. The dynamic handsplint is intended to be used for the rehabilitation of certain groups of muscles of a partially paralyzed hand in order to regain natural function and to prevent further disfigurement. The change in shape can be obtained by spanning each finger (except the thumb) of the glove with a number of strips which are interconnected under angles of 90° by rigid joints. To impose flexion one starts with straight strips. After mechanical deformation of the martensitic structure over a certain radius of curvature, the hand reaches an extension phase (Figure 18).

On raising the temperature above the transition temperature, the deformed strips remember their original, straight shape, to which they return, while the recovery force moves the hand. Because of the rigid, 90° connection between the strips, a flexion phase will be reached (Figure 19). On subsequent cooling below the transition temperature, the patient can easily straighten the hand since the resistance of the strips is low due to the low modulus of the shape memory material in its low temperature condition. A new cycle can then be started. To obtain an imposed extension one starts with bent strips which are initially deformed to straight strips. If such a shape memory effect system combined with an appropriate heating and cooling system can be fixed in a compact way to the glove, this dynamic handsplint offers practical, aesthetic, and psychological advantages for the patient, when compared with existing devices.

4. Heart and Vascular Surgery

In this discipline of medicine three applications of the shape memory effect of Nitinol have been reported: an artificial heart muscle,[74,75] a vena cava filter,[76] and an intracranial aneurysm clip.[77] The artificial heart muscle is the contractile element of an artificial heart chamber made of ethylene vinyl acetate copolymer. The artificial heart muscle consists of segments of Nitinol wires using a modified sinusoidal continuous wave form, anchored to

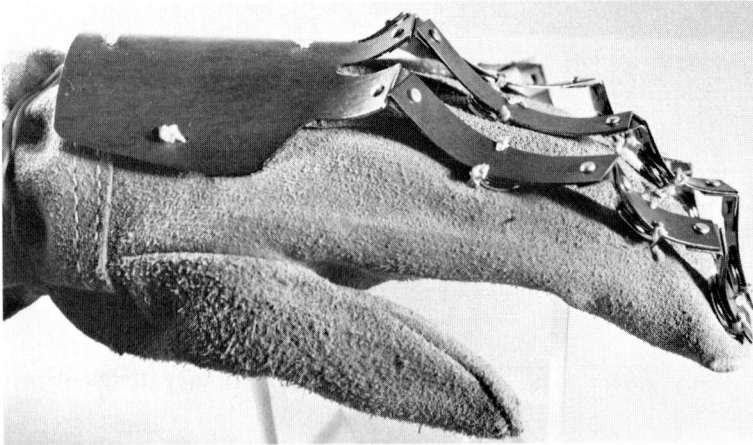

FIGURE 18.   Dynamic handsplint in extension phase.

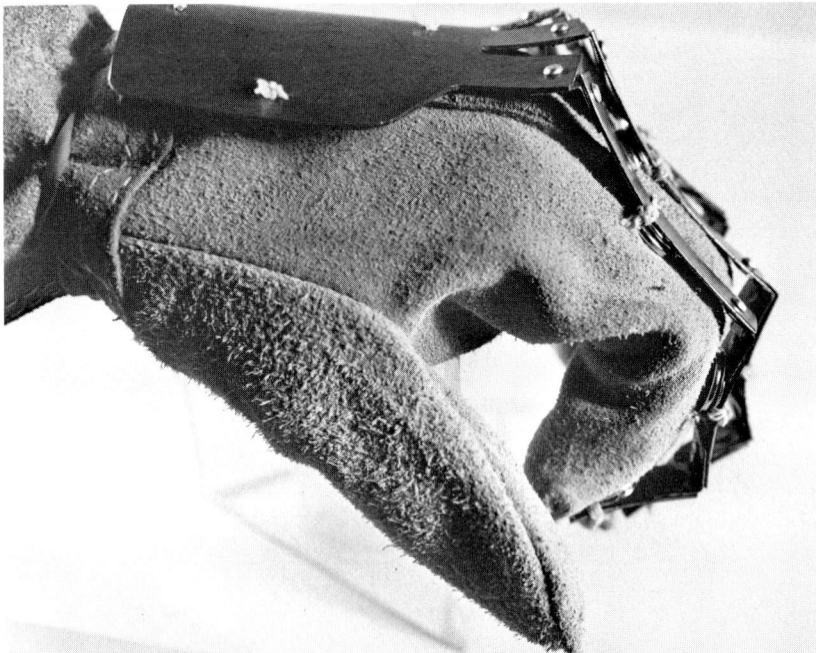

FIGURE 19.   Dynamic handsplint in flexion phase.

the exterior of the chamber in different contractile modules to attempt to replicate the contractility of the left heart ventricle. Upon cyclic heating of the different groups of the Nitinol wires above the transformation temperature range (which is selected to lie between 40 and 54°C) by an electrical energy source with a preset time-sequence, different portions of the heart chamber are contracted sequentially to produce pumping.

The rate of contraction of the different groups is related to the application and dissipation of heat, involving a limitation of the contractile rate for each group. Upon the reverse transformation on cooling, the wires will be stretched by the return of the elastic chamber wall to its initial diastolic orientation. An initial model has proven capable of ejecting a 25 to 35-cc water bolus up a 160-cm pressure head at a contractile rate of the heart chamber of 12 to 15 times per minute in vitro.

A vena cava filter using the shape memory effect has been evaluated by Simon et al.[76] With this device, acute pulmonary thromboembolism can be treated if the use of anticoagulants or thrombolytic drugs are precluded by the danger of internal bleeding. It is also a substitute for the surgical interruption of the vena cava or other great veins by ligation, stapling, clipping, or implantation of obstructive devices, such as miniature umbrellas or balloons. None of these surgical procedures is simple or without risk.

The blood clot filter is in its initial, martensitic configuration a straight Nitinol wire with a transformation temperature range well below body temperature. The chilled wire can be inserted into the patient's venous system with a standard angiographic catheter, while an infusion of cooled normal saline keeps the wire below its transformation temperature and hence straight during this procedure, which takes 20 to 30 sec.

Upon reaching the vena cava, the wire is extruded into the venous blood stream, whereupon it immediately transforms into its preset, complicated rigid filter shape. By using such devices, blood clots formed in the legs, pelvis, or thighs and subsequently dislodged are prevented from reaching the heart and lungs, while it permits the passage of normal blood. The filter has been designed to lock itself firmly into the wall of the vena cava at the chosen level. Experiments on dogs have been very encouraging, while for human applications the device promises greater safety (only a local anesthetic is required), simplicity, and speed of implantation.

At last an intracranial aneurysm clip to be used to tie off unwanted bulges in arteries has been designed by Netsu et al.[77] After experimenting with four different versions of aneurysm clips, it was found that a basic silver clip straddled by a supplementary clip made of Nitinol met all mechanical conditions for practical use and could be removed easily with only a local heating.

VI. CONCLUDING REMARKS

As shown, the shape memory alloys exhibit a number of remarkable mechanical properties, which open new possibilities in engineering in general and in biomedical engineering in particular. Until now, only one of the proposed applications in medicine has reached the commercial level, while the other ones are still in the experimental state. However, there is no doubt that if all metallurgical problems are solved and reliable devices can be produced, the shape memory alloys will become involved in the biomedical engineering.

ACKNOWLEDGMENTS

The author would like to thank Prof. Dr. ir. L. Delaey and ir. J. Van Humbeeck for many useful comments and discussions.

REFERENCES

1. **Ölander, A.,** *J. Am. Chem. Soc.,* 54, 3819, 1932; as cited in **Otsuka, K. and Wayman, C. M.,** Pseudo-elasticity and stress-induced martensitic transformations, *Int. Q. Sci. Rev. J.,* Reviews on the deformation behavior of materials, 2(2), 1977.
2. **Bush, R. E., Leudeman, R. T., and Gross, P. M.,** Alloys of Improved Elastic Properties, AMRA CR 65-02/1, AD 629726, U.S. Army Materials Research Agency, 1966.
3. **Greninger, A. B. and Mooradian, V. G.,** Strain transformation in metastable beta copper-zinc and beta copper-tin alloys, *Trans. AIME,* 128, 337, 1938.

4. **Kurdjumov, G. V. and Khandros, L. G.,** Dokl. Akad. Nauk. SSSR, 66, 211, 1949; (as cited in Reference 10).

5. **Chang, L. C. and Read, T. A.,** Plastic deformation and diffusionless phase changes in metals — the gold-cadmium beta phase, *Trans. AIME*, 191, 47, 1951.

6. **Buehler, W. J., Gilfrich, J. V., and Wiley, R. C.,** Effect of low-temperature phase changes on the mechanical properties of alloys near composition TiNi, *J. Appl. Phys.*, 34, 1475, 1963.

7. **Cross, W. B., Kariotis, A. H., and Stimler, F. J.,** Nitinol Characterisation Study, NASA CR-1433, National Aeronatics and Space Administration, Houston, 1969.

8. **Jackson, C. M., Wagner, H. J., and Wasilewski, R. J.,** 55-Nitinol, the Alloy with a Memory: Its Physical Metallurgy, Properties and Applications, NASA-SP 5110, National Aeronautics and Space Administration, Houston, 1972.

9. **Pops, H. and Ridley, N.,** Influence of aluminium on the martensitic transformation of beta phase Cu-Zn alloys, *Metall. Trans.*, 1, 2653, 1970.

10. **Delaey, L., Krishnan, R. V., Tas, H., and Warlimont, H.,** Review: thermoelasticity, pseudoelasticity and the memory effects associated with martensitic transformations, *J. Mater. Sci.*, 9, 1521, 1974.

11. **Delaey, L., Deruyttere, A., Aernoudt, E., and Roos, J.,** Shape Memory Effect, Superelasticity and Damping in Copper-Zinc-Aluminium Alloys, Report 78R1, Department Metaalkunde, Katholieke Universiteit Leuven, Belgium, 1978.

12. **Wang, F. E., Ed.,** Symposium on TiNi and Associated Compounds, NOLTR 68-16, U.S. Naval Ordnance Laboratory, White Oak, Silver Spring, Md. 1968.

13. **Perkins, J., Ed.,** *Shape Memory Effects in Alloys*, Plenum Press, New York, 1975.

14. **Cohen, M., Olsen, G. B., and Clapp, P. C.,** On the classification of displacive phase transformations, in *Proc. ICOMAT 79*, MIT Press, Cambridge, Mass. 1979, 1.

15. **Tas, H., Delaey, L., and Deruyttere, A.,** Stress-induced transformations and the shape memory effect, *J. Less-Common Met.*, 28, 141, 1972.

17. **Wasilewski, R. J.,** On the nature of the martensitic transformation, *Metall. Trans.*, 6A, 1405, 1975.

18. **Wayman, C. M. and Shimizu, K.,** The shape memory ("Marmem") effect in alloys, *Met. Sci. J.*, 6, 175, 1972.

19. **Wasilewski, R. J.,** The effects of applied stress on the martensitic transformation in TiNi, *Metall. Trans.*, 2, 2973, 1971.

20. **Wasilewski, R. J.,** Martensitic transformation and fatigue strength in TiNi, *Scr. Metall.*, 5, 207, 1971.

21. **Tas, H., Delaey, L., and Deruyttere, A.,** Stress induced phase transformations and the shape memory effect in β_1' Cu-Al martensite, *Scr. Metall.*, 5, 1117, 1971.

22. **Barceló, G., Rapacioli, R., and Ahlers, M.,** The rubber effect in Cu-Zn-Al martensite, *Scr. Metall.*, 12, 1069, 1978.

23. **Otsuka, K. and Shimizu, K.,** Stress and strain induced martensitic transformations, in *Proc. ICOMAT 79*, MIT Press, Cambridge, Mass., 1979, 607.

24. **Gilfrich, J. V.,** X-ray diffraction studies on the titanium-nickel system, in *Proc. 11th Annu. Conf. on Application of X-ray Analysis*, Vol. 6, Plenum Press, New York, 1963, 74.

25. **Zijderveld, J. A., de Lange, R. G., and Verbraak, C. A.,** La transformation martensique des alliages titane-nickel au voisinage de la composition equiatomique, *Mem. Sci. Rev. Met.*, 63, 885, 1966.

26. **de Lange, R. G. and Zijderveld, J. A.,** Shape memory effect and the martensitic transformation of TiNi, *J. Appl. Phys.*, 39, 2195, 1968.

27. **Wayman, C.M.,** On memory effects related to martensitic transformations and observations in β-brass and Fe₃Pt, *Scr. Metall.*, 5, 489, 1971.

28. **Otsuka, K. and Shimizu, K.,** On the crystallographic reversibility of martensitic transformations, *Scr. Metall.*, 11, 757, 1977.

29. **Sakamoto, H., Otsuka, K., and Shimizu, K.,** The heredity of martensitic transformations in ordered alloys, *Scr. Metall.*, 12, 1147, 1978.

30. **Perkins, J.,** Residual stresses and the origin of reversible (two-way) shape memory effects, *Scr. Metall.*, 8, 1469, 1974.

31. **Mohamed, H. A. and Washburn, J.,** On the mechanism of the shape memory effect in Ni-Ti alloy, *Met. Trans.*, 7A, 1041, 1976.

32. **Pops, H.,** Stress-induced pseudo-elasticity in ternary Cu-Zn based bate prime phase alloys, *Metall. Trans.*, 1, 251, 1970.

33. **Eisenwasser, J. D. and Brown, L. C.,** Pseudoelasticity and the strain-memory effect in Cu-Zn-Sn alloys, *Metall. Trans.*, 3, 1359, 1972.

34. **Olsen, G. B. and Cohen, M.,** Thermoelastic behaviour in martensitic transformations, *Scr. Metall.*, 9, 1247, 1975.

35. **Golestaneh, A. A.,** Martensitic phase transformation in shape memory alloys, in *Proc. ICOMAT 79*, MIT Press, Cambridge, Mass., 1979, 679.

36. **Mercier, O. and Melton, K. N.,** The influence of an anistropic elastic medium on the motion of dislocations: application to the martensitic transformation, *Scr. Metall.,* 10, 1075, 1976.

37. **Edwards, G. R., Perkins, J., and Johnson, J. M.,** Characterizing the shape memory effect potential of Ni-Ti alloys, *Scr. Metall.,* 9, 1167, 1975.

38. **Lieberman, D. S., Schmerling, M. A., and Karz, R. W.,** Ferroelastic "memory" and mechanical properties in gold cadmium, *in Shape Memory Effects in Alloys,* Perkins, J., Ed., Plenum Press, New York, 1975, 203.

39. **Dvorak, I. and Hawbolt, E. B.,** Transformational elasticity in a polycrystalline Cu-Zn-Sn alloy, *Metall. Trans.,* 6A, 95, 1975.

40. **Yang, N. Y. C., Laird, C., and Pope, D. P.,** The cyclic stress-strain response of polycrystalline, pseudoelastic Cu-14.5 wt. pct. Al-3 wt. pct. Ni alloy, *Metall. Trans.,* 8A, 955, 1977.

41. **Brown, L. C.,** The fatigue of pseudoelastic single crystals of β-CuAlNi, *Metall. Trans.,* 10A, 217, 1979.

42. **Delaey, L., Janssen, J., Van de Mosselaer, D., Dullenkopf, G., and Deruyttere, A.,** Fatigue properties of pseudoelastic Cu-Zn-Al alloys, *Scr. Metall.,* 12, 373, 1978.

43. **Janssen, J., Follon, M., and Delaey, L.,** The fatigue properties of superelastic Cu-Zn-Al alloys, in *Proc. 5th Int. Conf. on Strength of Metals and Alloys,* Aachen, Vol. 2, Haasen, P., Gerold, V., and Kostorz, G., Eds., Pergamon Press, Frankfurt, 1979, 1125.

44. **Melton, K. N. and Mercier, O.,** Fatigue life of Cu–Zn–Al alloys, *Scr. Metall.,* 13, 73, 1979.

45. **Wasilewski, R. J.,** On the "reversible shape memory effect" in martensitic transformation, *Scr. Metall.,* 9, 417, 1975.

46. **Melton, K. N. and Mercier, O.,** Fatigue of NiTi thermoelastic martensites, *Acta Metall.,* 27, 137, 1979.

47. **Melton, K. N. and Mercier, O.,** Fatigue of NiTi and Cu–Zn–Al shape memory alloys, in *Proc. 5th Int. Conf. on Strength of Metals and Alloys,* Aachen, Vol. 2, Haasen, P., Gerold, V., and Kostorz, G., Eds., Pergamon Press, Frankfurt, 1979, 1243.

48. **Melton, K. N. and Mercier, O.,** The effect of the martensitic phase transformation on the low cycle fatigue behaviour of polycrystalline Ni-Ti and Cu-Zn-Al alloys, *Mater. Sci. Eng.,* 40, 81, 1979.

49. **Saburi, T., and Nenno, S.,** Reversible shape memory in Cu-Zn-Ga, *Scr. Metall.,* 8, 1363, 1974.

50. **Banks, R.,** Nitinol heat engines, in *Shape Memory Effects in Alloys,* Perkins, J., Ed., Plenum Press, New York, 1975, 537.

51. **Cutright, D. E., Bhaskar, S. N., Perez, B., Johnson, R. M., and Cowan, G. S. M.,** Tissue reaction to Nitinol wire alloy, *Oral Surg.,* 35, 578, 1973.

52. **Castleman, L. S.,** Biocompatibility of Nitinol alloy as an implant material, in *Proc. 5th Annu. Int. Biomaterials Symp.,* Clemson University, Clemson, S.C., 1973.

53. **Castleman, L. S., Motzkin, S. M., Alicandri, F. P., Bonawit, V. L., and Johnson, A. A.,** Biocompatibility of Nitinol alloy as an implant material, *J. Biomed. Mater. Res.,* 10, 695, 1976.

54. **Motzkin, S. M., Castleman, L. S., Szablowski, W., Bonawit, V. L., Alicandri, F. P., and Johnson, A. A.,** Evaluation of Nitinol compatibility by cell culture, in *Proc. 4th New England Bioengineering Conf.,* Yale University, New Haven, Conn., 1976, 301.

55. **Hughes, J. L.,** Evaluation of Nitinol for Use as a Material in the Construction of Orthopaedic Implants, DAMD 17-74-C-4041, U.S. Army Medical Research and Development Command, Fort Detrick, Frederick, Md., 1977.

56. **Schmerling, M. A., Wilkov, M. A., Sanders, A. E., and Woosley, J. E.,** Using the shape recovery of Nitinol in the Harrington rod treatment of scoliosis, *J. Biomed. Mater. Res.,* 10, 879, 1976.

57. **Haasters, J., Bensmann, G., and Baumgart, F.,** Memory alloys - new material for implantation in orthopedic surgery, part 2 in *Current Concepts of Internal Fixation of Fractures,* Uhthoff, H. K., Ed., Springer-Verlag, New York, 1980, 128.

58. **McDonald Schetky, L. and Sims, R. B.,** Design concepts for actuators using shape memory effect brasses, in *Proc. ICOMAT 79,* MIT Press, Cambridge, Mass., 1979, 693.

59. **Powley, D. G. and Brook, G. B.,** The design and testing of a memory metal actuated boom release mechanism, in Proc. 12th Aerospace Mechanisms Symp., NASA CP-2080, Ames Research Center, Moffet Field, Calif., 1979, 119.

60. **Wagner, H. J. and Jackson, C. M.,** What you can do with that "memory" alloy, *Mater. Eng.,* 70, 28, 1969.

61. **Heisterkamp, C. A., Buehler, W. J., and Wang, F. E.,** 55-Nitinol — a new biomaterial, paper presented at the 8th Int. Conf. on Medical and Biomedical Engineering, Chicago, July 1969.

62. **Fannon, R. D., Lower, B. R., and Laufe, L. E.,** U.S. Patent 3,620,212, 1971.

63. **Wilson, B. C.,** U.S. Patent 3,890,977, 1975.

64. **Andreasen, G. F.,** U.S. Patent 4,037,324, 1977.

65. **Andreasen, G. F.,** A clinical trial of alignment of teeth using a 0.019 inch thermal nitinol wire with a transition temperature range between 31°C and 45°C, *Am. J. Orthod.,* 78, 528, 1980.

66. **Andreasen, G. F. and Morrow, R. E.,** Laboratory and clinical analyses of Nitinol wire, *Am. J. Orthod.,* 73, 142, 1978.
67. **Civjan, S., Huget, E. F., and DeSimon, L. B.,** Potential applications of certain nickel-titanium (Nitinol) alloys, *J. Dent. Res.,* 54, 89, 1975.
68. **Bensmann, G., Baumgart, F., Hartwig, J., and Haasters, J.,** Untersuchungen der Memory-Legierung Nickel-Titan und Überlegungen zu ihrer Anwendungen im Bereich der Medizin, *Tech. Mitt. Krupp Forschungsber.,* 37, 21, 1979.
69. **Johnson, A. A. and Alicandri, F. P.,** U.S. Patent 3,786,806, 1974.
70. **Baumgart, F., Bensmann, G., and Hartwig, J.,** Mechanische Probleme bei der Nutzung des Memory-Effektes für Osteosyntheseplatten, *Tech. Mitt. Krupp Forschungsber.,* 35, 157, 1977.
71. **Kousbroek, R., Van der Perre, G., Aernoudt, E., and Mulier, J. C.,** Shape memory effect in biomedical devices, in *Advances in Biomaterials,* Vol. 3, Winter, G. D., Gibbons, D. F., and Plenk, H., Eds., John Wiley & Sons, New York, 1982, 767.
72. **Kousbroek, R.,** Ph.D-thesis, Katholieke Universiteit, Leuven, Belgium, 1983.
73. **Baumgart, F., Bensmann, G., Haasters, J., Nölker, A., and Schlegel, K. F.,** Zur Dwyerschen Skoliosenoperation mittels Drähten aus Memory-Legieurungen, *Arch. Orth. Traum. Surg.,* 91, 67, 1978.
74. **Sawyer, P. N., Page, M., Rudewald, B., Lagergren, H., Baselius, L., McCool, C., Halperin, W., and Srinivasan, S.,** Characteristics of the human heart: design requirements for replacement, *Trans. Am. Soc. Artif. Int. Organs,* 17, 470, 1971.
75. **Page, M. and Sawyer, P. N.,** U.S. Patent 3,827,426, 1974.
76. **Simon, M., Kaplow, R., Salzman, E., and Freiman, D.,** A vena cava filter using thermal shape memory alloy, *Radiology,* 125, 89, 1977.
77. **Netsu, N., Iwabuchi, T., and Honma, T.,** A study of the TiNi intracranial aneurysm clip having the shape memory effect, *Bull. Res. Inst. Min. Dress. Met. Tohoku Univ.,* 34, 67, 1978.

Chapter 4

CORROSION

R. W. Lycett and A. N. Hughes

TABLE OF CONTENTS

I. THE CORROSION PROCESS

Corrosion may be regarded as the usually unwanted interaction of a metallic component with the environment in which it exists. During the process, metal ions may be lost from the metal surface to form either a solid corrosion product or one that is soluble in its environment. The corrosion environment may in general be liquid or gaseous. However, since this chapter relates to the corrosion of orthopedic implants, emphasis will be given to aqueous corrosion. It is often easy to overlook the fact that the body fluids, in which orthopedic devices are expected to function, are essentially a 1w/o saline solution held at a temperature of 37°C and create a very aggressive environment in terms of metallic corrosion. However, the experience gained in the use of metals in the body, particularly since the mid 1930s, has led to the adoption of various classes of metals which offer a very high resistance to corrosion in the body; those most commonly employed are based on the 18% chromium — 10% nickel stainless steels, certain cobalt base alloys, and alloys based on titanium.

With the exception of certain noble metals, which may occur in the elementary state, metals are usually found in nature as compounds, principally oxides and sulfides. The extraction process by which these metals are reduced from their ores only occurs with an increase in free energy and hence requires an external supply of energy, as, for example, in the electrical reduction of aluminum oxide to aluminum. Since most reactions of the type metal $\rightarrow$ metal compound occur with a decrease in free energy, after purification and fabrication, metals may react spontaneously with many liquids or gases. It is this chemical reaction between a metal and an environment to form a compound which constitutes corrosion and may be regarded as the reversion of the metal to a more "natural" state, i.e., the reverse of the extraction process. Corrosion reactions are chemical processes that take place at the surface of the metal and obey well-established laws.

It has been established that corrosion in aqueous media is an electrochemical process. It is necessary, therefore, to discuss in some detail the electrochemical principles most relevant to corrosion processes.

Two kinds of chemical reactions usually occur on a metal surface exposed to an aqueous solution. There is an oxidation or anodic reaction, which produces a supply of electrons of the type:

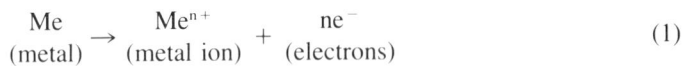

$$\underset{\text{(metal)}}{\text{Me}} \rightarrow \underset{\text{(metal ion)}}{\text{Me}^{n+}} + \underset{\text{(electrons)}}{n\text{e}^-} \qquad (1)$$

and a reduction or cathodic reaction which consumes the electrons produced by the anodic reaction. The reduction of dissolved oxygen and the production of hydrogen ions are the two principal cathodic reactions, represented as:

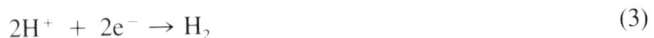

$$O_2 + 4e^- + 2H_2O \rightarrow 4OH^- \qquad (2)$$

$$2H^+ + 2e^- \rightarrow H_2 \qquad (3)$$

The latter reaction is favored in acid solutions, but is nevertheless relevant to body fluids since the normally neutral pH values of the bulk fluids can become acidic when the supply of oxygen is restricted either by geometric or physiological conditions.

As with any electrochemical cell, both the anodic and cathodic reactions must occur simultaneously and, to preserve electrical neutrality, must proceed at the same rate. The sites of these reactions, i.e., the anode and the cathode, need not be adjacent, but can be widely separated, providing they remain in electrical contact. The conditions governing the positions of anodes and cathodes, which under certain circumstances can be mobile, depend

Table 1
STANDARD ELECTRODE
POTENTIALS AT 25°C

Electrode reaction			$E°(V)$
Au	$= Au^{3+}$	$+ 3e^-$	$+ 1.5$
Ag	$= Ag^+$	$+ e^-$	$+ 0.799$
Fe^{2+}	$= Fe^{3+}$	$+ e^-$	$+ 0.771$
Cu	$= Cu^{2+}$	$+ 2e^-$	$+ 0.337$
H_2	$= 2H^+$	$+ 2e^-$	
Fe	$= Fe^{3+}$	$+ 3e^-$	$- 0.036$
Ni	$= Ni^{2+}$	$+ 2e^-$	$- 0.250$
Fe	$= Fe^{2+}$	$+ 2e^-$	$- 0.440$
Cr	$= Cr^{3+}$	$+ 3e^-$	$- 0.744$
Zn	$= Zn^{2+}$	$+ 2e^-$	$- 0.763$
Al	$= Al^{3+}$	$+ 3e$	$- 1.66$

Note: Volts vs. normal hydrogen electrode.

From Latimer, W. M., *Oxidation Potentials,* 2nd ed., Prentice-Hall, Englewood Cliffs, N.J. 1952. With permission.

upon both the metal surface and the environment. Variation in the oxygen concentration, for example, can lead to the establishment of anodes in regions of low oxygen concentration. The anodic sites may cover only a small percentage of the surface, the remainder providing a large cathodic area. Under these conditions the anodes can be very active leading to pitting. If two metals are in contact, one may form the anode, and the other may form the cathode, leading to preferential attack of only one material. Variations in the homogeneity of the metal surface caused by the presence of grain boundaries, nonmetallic inclusions, etc. can again lead to the formation of anodic and cathodic sites. These different situations which result in the many varied forms of corrosive attack observed in orthopedic implants are discussed in greater detail later in the chapter.

Metal loss is the immediate consequence of the anodic oxidation process. The electron flow between the anode and the cathode constitutes the corrosion current, the value of which is determined by the rate of production of electrons and hence the rate of corrosion.

For electrons to flow between the anodes and cathodes, a driving force must exist. This is the difference in electrical potential between the anodic and cathodic sites. The difference exists because each oxidation and reduction reaction has associated with it a potential, determined by the tendency for the reaction to occur spontaneously. Although the potential established between a metal and a solution cannot be measured in absolute terms, the potential difference between the metal and another electrode can be determined. Changes in potential difference can then be related to the metal under investigation providing the potential of the other electrode remains constant and consequently acts as a reference. (Several reference electrodes can be chosen, if the current drawn from them is small, in the order of 10^{-12} A cm^{-2}.)

A metal in contact with a solution containing its metal ions at unit activity establishes a fixed potential difference with respect to every other metal in the same condition. The list of these potentials comprises the electromotive series which is given in Table 1. The potentials given in the table are differences between the metal and the hydrogen reference electrode, which is arbitrarily chosen as 0.00 V and 25°C.

While Table 1 is frequently very useful, the potentials obtained will differ considerably

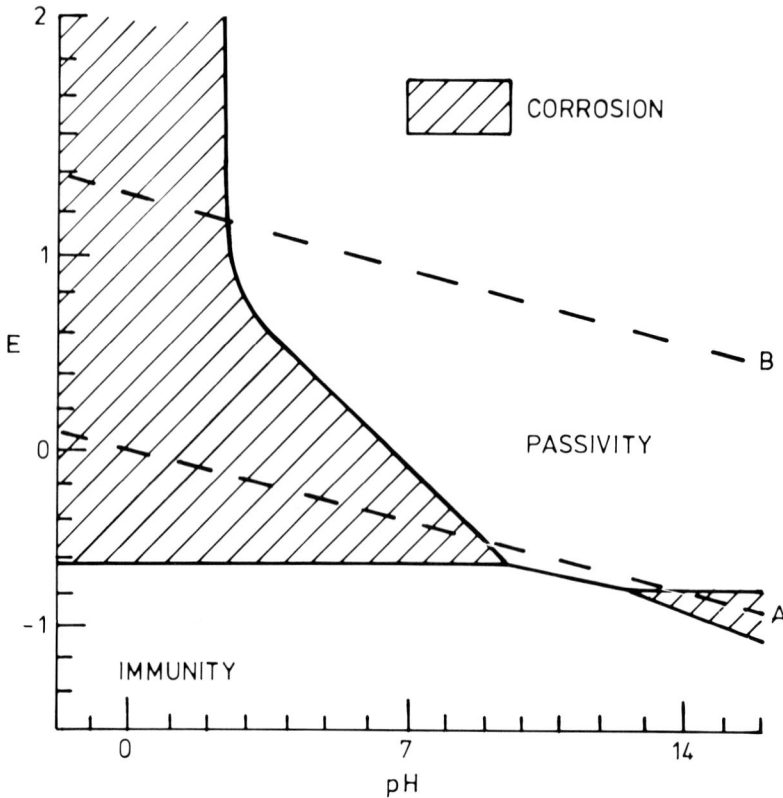

FIGURE 1.   Simplified Pourbaix diagram for iron showing the gaseous evolution
lines A and B.

if the metal surface is bare or if it is covered by an oxide film or is contaminated (by, for example, ions of chlorine or sulfur). All surfaces have an energy associated with them and the material will spontaneously try to lower this surface energy. One method it may adopt is the adsorption of compounds from the environment, notably oxygen to form a surface oxide.

A corrosion resistant (protective) film may result if, after initial exposure to the environment a thin, uniform, adherent, continuous film is formed on the metal surface. Under these circumstances, the rate of growth will tend to zero quickly. The film must be coherent with the metal surface, and the strain in the layer must be insufficient to rupture it. Regeneration of the film must also be possible if breaks occur. If the resulting oxide film is not protective, continuous loss of metal occurs with subsequent loss of mechanical strength of the component. This is a common feature, most notably with ferrous structures, the rate of attack being increased, with a subsequent reduction in the lifetime of the component, if the atmosphere contains appreciable amounts of chloride ions, sulfur, or SO_2. (Any degree of protection offered by the oxide film is greatly dependent upon the particular environment, especially if it is strongly oxidizing or reducing.)

A useful method of studying the relation of potential to corrosion is the Pourbaix diagram which plots potential against pH (the log of the hydrogen ion concentration and a measure of the acidity of a solution). A simplified Pourbaix diagram for iron is given in Figure 1.

By measuring potential and pH, a Pourbaix diagram can be used to determine whether a metal surface is in a region where the tendency for corrosion is high, a region where corrosion tendency is nil (immunity), and a region where although there exists a propensity for corrosion, there is also a tendency for a protective oxide film to form. This protective or passive film, while not removing the tendency for corrosion, does reduce the rate of attack for many practical purposes to zero. (Passivity is discussed in greater detail later in the chapter.) The lines A and B in Figure 1 represent the variation of potential with pH for the hydrogen and oxygen evolution reactions. Above the oxygen line (B), oxygen is evolved, while below the hydrogen line (A), hydrogen evolution occurs. In practice it is usually necessary to take the potential some way above line B and below line A for gaseous evolution to occur. This overpotential is determined by the properties of surface films and of the metal surface and will vary with material and environment. Using the Pourbaix diagram, potential and pH measurement can therefore give some indication as to the likely behavior of a prospective implant material, whether towards immunity, passivity, or corrosive attack.

II. KINETICS OF CORROSION

While a knowledge of the corrosion tendencies of a metal as shown by, for example, a Pourbaix diagram is a useful indication of its likely behavior as an orthopedic implant, a more quantitative description of the process is necessary to accurately predict the performance in the body. Corrosion rate kinetics provide the information describing the number of metal ions released in a given period of time, which in turn can be related to the effects (toxicity) of those ions, in the body. Obviously higher corrosion rates can be tolerated from a metal producing ions of low toxicity than that from a metal producing more harmful ions. The converse is also true, in that providing a metal has a very low corrosion rate, the relatively high toxicity of its ions may be tolerated.

Corrosion reactions must follow the electrochemical law that the rate of the anodic reaction must equal that of the cathodic reaction. It follows, therefore, that if either of the reactions can be retarded, then the rate of corrosion can be reduced. This can be achieved by a process known as polarization, which results in a change in the potential of the metal surface and may occur due to processes occurring at the metal surface, for example, film formation or ion absorption (activation polarization) or by the concentration of metal ions in the environment (concentration polarization). Polarization in a positive direction with an increase in potential is termed anodic polarization, while the reverse is cathodic polarization. The degree of polarization is a measure of how much the anodic and cathodic reactions are affected by the processes mentioned above.

When a current flows between an anode and a cathode, the potential of both will become polarized, thereby reducing the potential difference between the anodic and cathodic areas. This process is shown schematically in Figure 2. At equilibrium, the metal will have a corrosion potential of E_{corr} and a corrosion current density between the anode and cathode areas of i_{corr}.

As the degree of polarization increases, then the rate of corrosion falls and Figure 3 shows the effects of anodic polarization and cathodic polarization (hydrogen overpotential) on the values of the corrosion rate i_{corr}. The change in potential with current flow may vary, and Figure 4 illustrates the different effects observed. In Figure 4 section 1, it can be seen that changes in the cathodic polarization curve have only a minor effect upon i_{corr} compared with the effects of the anodic curve. This is an example of the anodic reaction determining the rate of corrosion and is denoted therefore as anodic control. Figure 4 part 2 shows the reverse situation, cathodic control, where only changes in the cathodic curve significantly affect i_{corr}. In Figure 4 part 3 polarization of both curves will have a similar effect upon i_{corr} and hence this is an example of mixed control. Curves of the type shown in Figure 4 are called Evans diagrams and are used extensively in corrosion science to determine corrosion rates and to study which reactions control the overall process.

Polarization of samples in the laboratory may be achieved using apparatus of the type

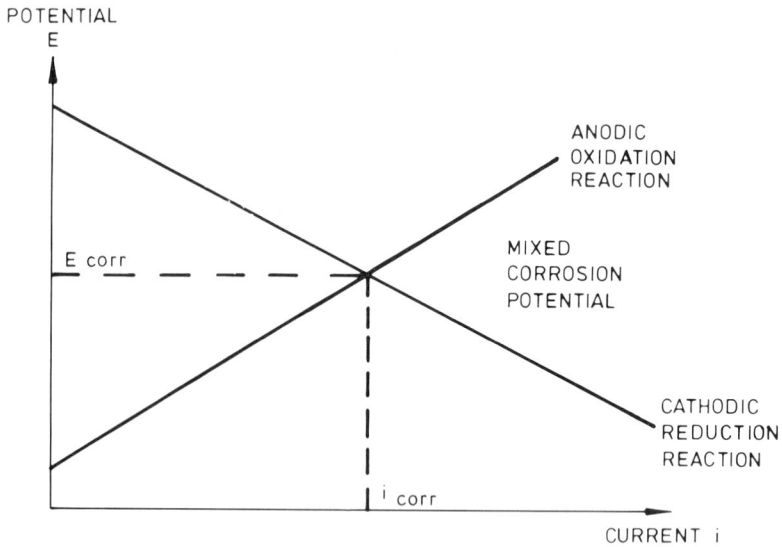

FIGURE 2. Schematic representation of polarization of anodic and cathodic reactions in a low resistance medium.

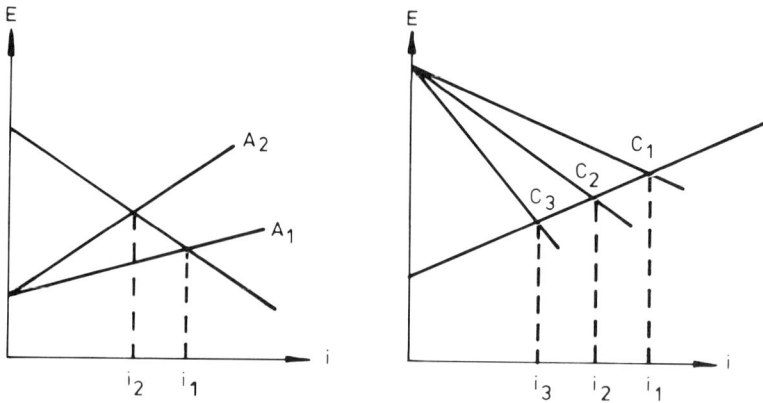

FIGURE 3. The effects of anodic and cathodic polarization on the corrosion rate.

shown schematically in Figure 5. The battery supplies a flow of electrons via the auxiliary electrode, frequently made of platinum, and causes the potential on the specimens to be changed from the open circuit or rest potential. The potential of the sample can be determined, with respect to the reference electrode using a millivoltmeter. The current required to maintain the sample at the new potential is monitored using an ammeter. The polarization curves are measured by applying a series of potentials to the sample and measuring the current supplied by the auxiliary electrode. An example of the type of curve obtained is shown in Figure 6. When the potential is plotted as a function of the current density, part of the curve is usually linear and it is possible to extrapolate the linear position back to obtain true polarization curves. The deviations from the true curves occur because at potentials close to E_{corr}, the specimen surface will have both anodic and cathodic regions on the surface and hence the measured curve will be a combination of the true anodic and cathodic curves. Further extrapolation of the true curves can be made back to the reversible potentials of the anodic and cathodic lines, where the metal is in equilibrium with its ions. Figure 6B shows the results recorded for polarization curves of iron in 0.52 N sulfuric acid.

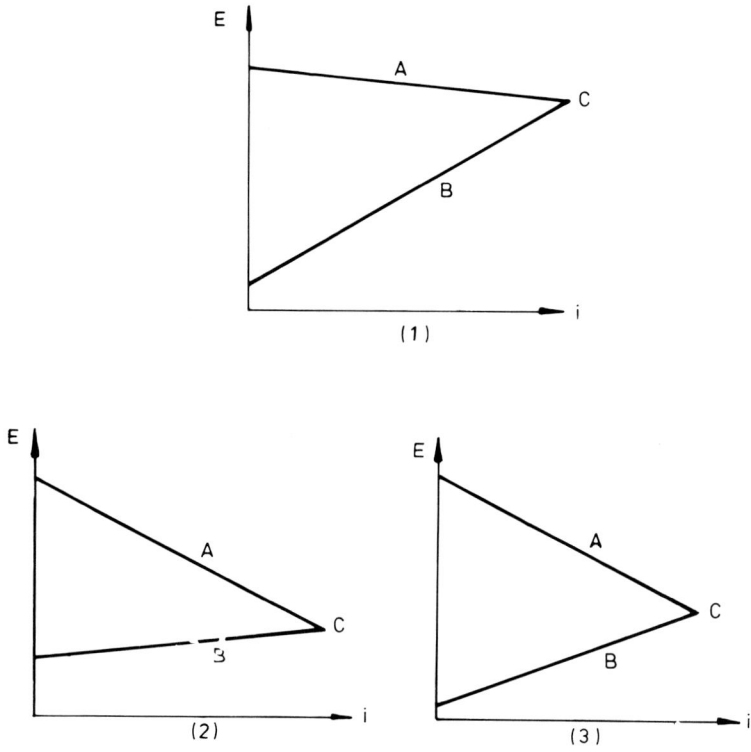

FIGURE 4. The effect upon the corrosion reaction of different levels of polarization (A = cathodic reduction, B = anodic oxidation, C = mixed corrosion potential).

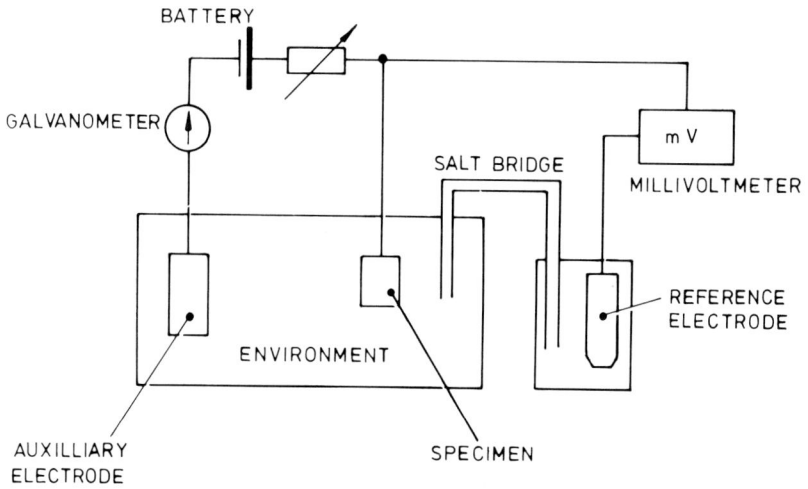

FIGURE 5. Apparatus used for the measurement of corrosion rate.

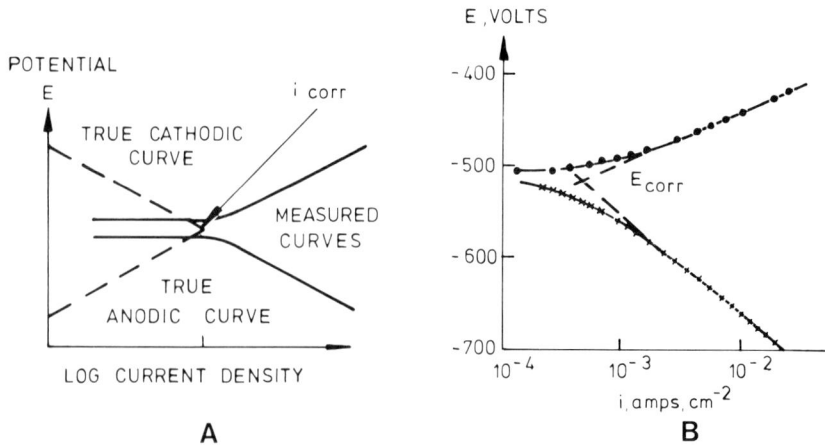

FIGURE 6.   (A) Schematic polarization curve; (B) polarization behavior of iron in 0.52 *N* sulfuric acid. (From Makrides, A. C., *J. Electrochem. Soc.*, 104, 869, 1960. With permission.)

III. PASSIVITY

The preceding discussion has illustrated the importance of both corrosion tendency (a thermodynamic property) and corrosion rate (a kinetic property) in determining the amount of corrosive attack occurring at a metal surface. The further important factor in determining the rate of corrosion is the phenomenon of passivity. This is especially true for surgical implants where almost all of the metals and alloys used exhibit passivity under certain conditions.

Passivity is the presence of an oxide film on the surface of the material which markedly lowers the corrosion rate to a value close to zero, despite any thermodynamic predictions indicating a tendency for a much more aggressive attack. Figure 7 illustrates schematically the polarization curve for a material exhibiting passivity. At low potentials, increasing the potential increases the corrosion current. At point B, however, a critical current density $I_{critical}$ is exceeded and the current density falls along the path BC to point D. BCD is an unstable region where the corrosion rate cannot be maintained, the current falling rapidly to D. The potential here is referred to as E_{pp}, the primary passive potential. The fall in current density may be some orders of magnitude and corresponds to the existence of a thin stable oxide film of comparatively high electrical resistance. Although very stable, the film slowly dissolves, and a small current is required to maintain it.

Passivity is difficult to establish and maintain in the presence of chloride ions, and increasing concentrations of chloride ions lead to an increase in the critical current density required to establish passivity, resulting in a shift of the section DE in Figure 7 to higher current density levels. Further increases in the potential above E lead to a region where pitting occurs, or alternatively a region of transpassivity may exist where oxygen evolution and possibly an increase in the corrosion rate may occur.

The corrosion of metal surface is dependent upon both the associated anodic and cathodic processes, and the corrosion potential of the metal is given at the intersection of the anodic and cathodic polarization curves. Therefore although an alloy may be capable of exhibiting passivity, it can only be used in the passive condition, if the mixed corrosion potential E_{corr} occurs in the passive region DE as shown in Figure 8, curve B. Under these conditions, the alloy is described as self-passivating (no externally applied current being necessary for the material to exhibit passivity).

In Case A of Figure 8, pitting will occur. Case D is indicative of reducing conditions and the resulting corrosion rate may be high. In Case C there are three points of intersection in

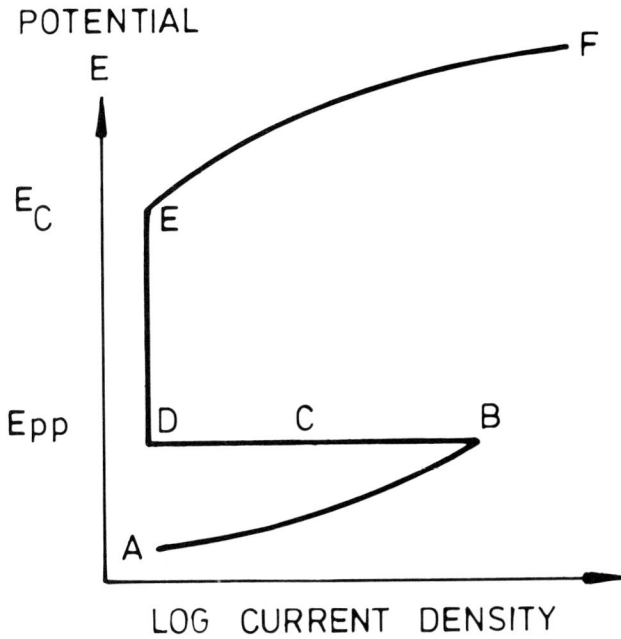

FIGURE 7.   Polarization curve for a material exhibiting passivity.

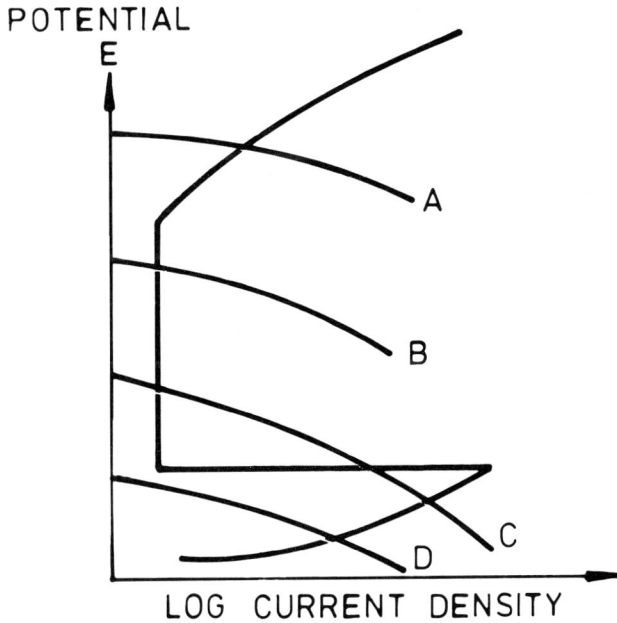

FIGURE 8.   Mixed corrosion potentials produced by different cathodic reaction curves.

the active, passive, and unstable regions. Here any passivity which is exhibited will be metastable and the surface will vary between active and passive states. The condition for passivity may be stated that at E_{pp}, $I_{cathodic}$ must be greater than $I_{critical}$. The methods of making an alloy passive can be to increase $I_{cathodic}$ at E_{pp} (as with Ti-Pd alloys) or by lowering

$I_{critical}$ (as with the addition of Cr to iron alloys). Other than alloying, the methods used to make alloys passive are

1. Application of a current using apparatus of the type shown in Figure 5
2. Modification of the environment (by the addition of chromate ions, for example)
3. By treating the material in an oxidizing solution, before exposure to the corrosive environment, to ensure the presence of a good passive film, for example, stainless steel treated in nitric acid

In the above discussion, some of the fundamentals of corrosion have been outlined. The effects of corrosion can be very serious in industrial applications due to loss of mechanical strength or perforation of tanks due to pitting. For surgical implant applications, however, a further consideration is patient sensitivity to the corrosion products. As discussed earlier, in aqueous corrosion, metal ions pass into the surrounding solution. Metal salts (and possibly wear products) will therefore be present in the region surrounding an implant. If crystalline particles of metal are present in large amounts, there is frequent evidence of giant cell reaction and of cell and tissue death. This is also particularly marked in the presence of cobalt which is believed to exist as soluble salts in solution. The process of giant cell formation and tissue and bone necrosis can lead to pain and loosening of the implant which is detrimental to its performance, leading to, for example, corrosion fatigue failure in hip joint prostheses.

Tissue sensitivity induced by metal salts can also lead to loosening, and chromium, cobalt, and nickel are all known to produce skin sensitivity. Patients with metal sensitivity may experience exaggerated tissue response to the wear and corrosion products. This results in a higher incidence of loosening in patients who experience tissue sensitivity.

It is clear from these findings that it is desirable to restrict the rate of corrosion occurring in surgical implants as this has the added benefit of reducing the probability of pain and loosening as well as maintaining the integrity and stress bearing capacity of the component.

IV. LOCALIZED CORROSION

Under conditions where the anodic and cathodic sites are mobile or randomly distributed, the resulting attack may be general over the whole surface. In many instances, however, the attack is concentrated at specific sites, resulting in many different forms of localized corrosive attack, the most common of which are described below.

A. Galvanic Corrosion

If two dissimilar metals placed in contact with each other are subsequently exposed to a conductive solution, an electrical potential will exist between them. This will serve as the driving force for the flow of current, with subsequent corrosion, referred to as galvanic corrosion, of one of the metals in the couple. The larger the potential difference between the two, the greater is the probability of corrosion of the less noble metal (galvanic attack). Galvanic corrosion only causes accelerated deterioration of the less noble material, which would have undergone attack even if placed in the solution in isolation.

Reference may be made to the electrochemical series to determine which component of a galvanic couple will be protected and which will be attacked. However, since the solution will not contain ions of unit activity and the metal surface may be covered by contaminants, reference to the series should be made with great care, as the idealized relative potential may be reversed in practice and the apparently more noble material attacked.

The phenomenon is clearly of importance in multicomponent devices, for example, bone plates and screws, where use of items of dissimilar materials may induce galvanic attack. The situation is frequently more complex, however, and the presence of minute particles from cutting tools used in specimen manufacture and instruments used to prepare patients for surgery may also lead to galvanic attack. Consequently all drill bits and screwdriver heads should wherever possible be of the same material as the proposed implant or of a material which will not disturb the passivity of the implant.

Heavily deformed and cold worked material, for example, swarf or metal turnings, may also result in galvanic attack when placed in contact with the same material which has been subjected to less cold work. Because of this it is essential to remove all swarf from the threads of screws to reduce the likelihood of loosening as a result of galvanic corrosion effects.

Components comprising of more than one material may be safely used if the potential of the less noble material is not raised from the region of passivity (DE in Figure 7). Such devices are successfully utilized in hip joint prostheses, for example, combining a material of good wear resistance for the head of the prosthesis with one of good fatigue resistance for the stem. Predictions of the behavior of double alloy implants are, however, more difficult than for single metal components and wherever possible such risks should be eliminated by using only a single material.

B. Pitting Corrosion

Pitting is a form of highly localized attack frequently occurring on a flat surface and resulting in the formation of holes on an otherwise unattacked surface. The shape of the pits varies from hemispherical to a wide variety of irregular shapes caused by undercutting with the result that the surface diameter may be much less than the diameter below the surface.

This form of attack is associated with the breakdown of a surface oxide film at discrete sites and is believed to occur under conditions intermediate between immunity and general corrosion. Favored sites for pit initiation are discontinuities in the oxide film, disrupted on an atomic scale by local deformation, for example, emergent screw dislocations, along emergent slip lines or at sheared edges. Alternative sites are points of compositional variations, for example, inclusions and phase or grain boundaries or places where there are variations in the environment composition. Scratches may also be favored nucleation sites.

At initiation, a high local concentration of metal ions will be produced by dissolution and continued dissolution will occur if the surface fails to repassivate. The metal ions thus produced may then be precipitated as a solid corrosion product often covering the mouth of the pit. This restricts the flow of ions into or out of the pit and allows the build-up of hydrogen ion concentration within the pit. To maintain charge neutrality and assuming a saline environment, the chloride ion concentration also increases leading to the formation of a highly acidic, high metallic ion concentration solution within the pit. Repassivation is even more difficult in these aggressive solutions leading to further pit propagation in a self-accelerating process. Figure 9 shows schematically the initiation and growth of a corrosion pit.

The anode/cathode ratio (pit area/unattacked metal surface) is very low resulting in high anodic current densities and rapid pit penetration. The initiation of the process is, however, frequently very slow with a long incubation period. The avoidance of rapidly growing pits requires that the material has its rest potential below the critical potential for pitting (E_c in Figure 7). Work on pitting propensity in simulated physiological saline solutions has shown that for titanium and chromium cobalt alloys, the probability of pit formation is very small. With stainless steel, however, there is a significant probability of pit formation. This is borne out in practice where the observation of pitting is confined to stainless steel components.

C. Crevice Corrosion

As the principal cathode reaction in aqueous solutions is the reduction of oxygen to

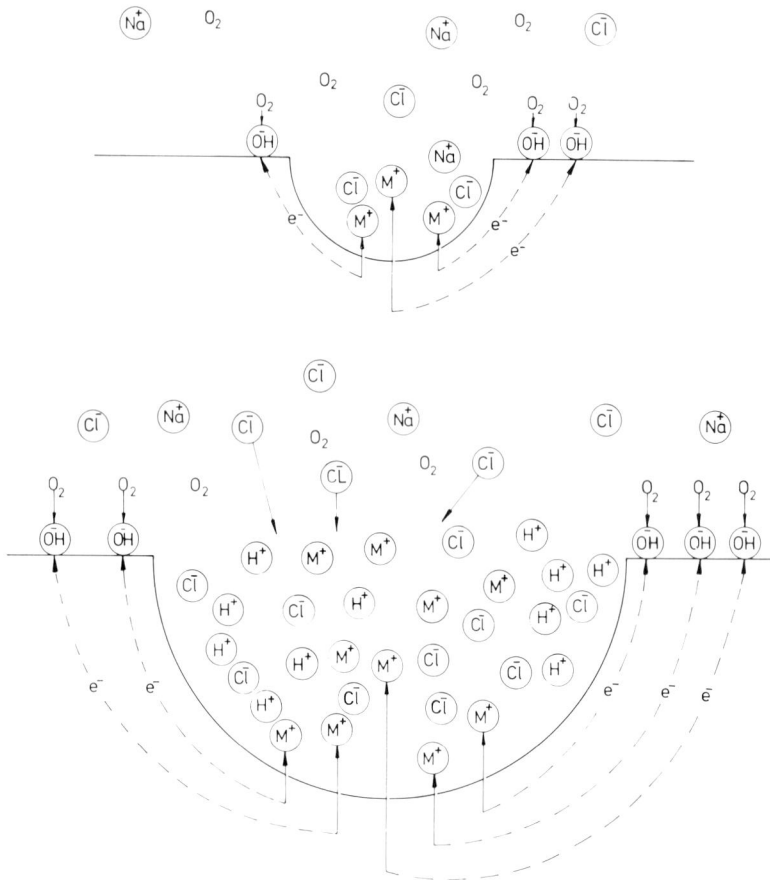

FIGURE 9.   Initiation and growth of pits in a sodium chloride environment.

hydroxyl ions, it follows therefore that regions of highest oxygen concentration and easy oxygen access will be principally cathodic, whereas those regions with restricted oxygen access will become anodic. A circuit will flow between a well-aerated electrode and a less-aerated one with the latter suffering preferential attack. This phenomenon is known as differential aeration and is important in many corrosion phenomena, noticeably crevice attack.

Because the electrolyte in a crevice has only limited access to the surrounding saline fluid, corrosion leads to a build-up of hydrogen and metal ions in the solution, together with a decrease in oxygen. The highly mobile chloride ion migrates into the crevice, and by a similar mechanism to that observed with pitting, the attack becomes accelerated. There is much similarity between the processes of crevice attack and pitting, and both may be described under the general heading of occluded cell corrosion.

As crevice attack occurs at interfaces, multicomponent devices, notably bone plates and screws, suffer from this type of attack. The recess beneath the screw head is a favored site for attack, particularly on stainless steel components, and the likelihood of occurrence is increased if the screws are not inserted and seated in the recesss correctly.

D. Grain Boundary Attack

All metals crystallize from the molten state to form thousands of grains (regions of regular atomic arrangements). Where grains meet there is a region of mismatch, the grain boundary.

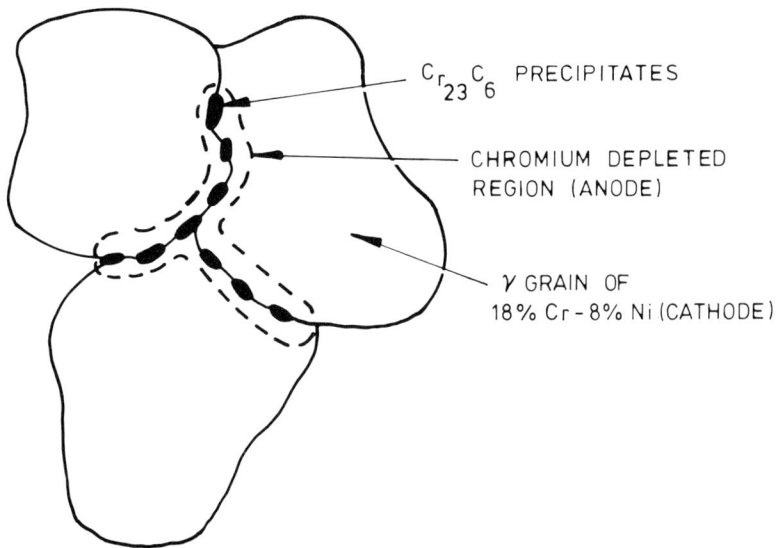

FIGURE 10. Carbide precipitates at grain boundaries leading to the setting up of an electrochemical cell and grain boundary dissolution.

These regions are less ordered and possess higher energy than the body of the grains and may suffer preferential attack, resulting in intergranular corrosion. The attack may be due to the higher energy of the boundaries or to alloy segregation, resulting in a potential drop between the boundary and the body of the grain sufficient to lead to the development of corrosion currents of considerable magnitude.

Stainless steels made of 18% Cr-8% Ni are susceptible to intergranular corrosion when heat treated in the temperature range 500 to 800°, as can occur, for example, in regions adjacent to welds or as a result of insufficient thermal control during forging. In this temperature range, chromium carbides $Cr_{23} C_6$ are precipitated at the grain boundaries and the immediately adjacent areas become depleted in chromium, as shown in Figure 10. These chromium depleted regions become active with respect to the body of the grain, and in the presence of an electrolyte can lead to the formation of an electrochemical cell. The low anode/cathode area ratio leads to corrosion under anodic control and the possibility of very severe attack.

This phenomenon can be remedied by the addition of strong carbide formers to the steel, for example, niobium or titanium. These will then form carbides preferentially to $Cr_{23} C_6$, thus preventing the formation of the chromium depleted region. Steel in this condition is said to be stabilized. Low carbon grades of steel, for example, AISI 316L also reduce the problem by limiting the carbon available to form carbide. Phosphorous and sulfur segregation have also been proposed as being responsible for weld decay, i.e., attack close to welds.

V. ENVIRONMENT ASSISTED CRACKING

Many alloys when subjected to the conjoint action of tensile stress and a corrosive solution will suffer cracking and premature failure. The process is a synergistic action since in many cases both the stress level and the solution themselves are relatively harmless and only induce premature failure when acting together. Three basic examples of the phenomenon will be considered here, stress corrosion cracking (SCC), corrosion fatigue, and fretting corrosion.

A. Stress Corrosion Cracking

The following general points can be made relating to SCC:

1.   Its occurrence is confined to specific alloy environment combinations, although the range of solutions which can cause cracking is increasing. The phenomenon is confined to alloys and does not occur in metals, although even "ultrapure" copper and iron have been observed to suffer from SCC ascribed to impurities at grain boundaries.
2.   The stress, which must have a tensile component, and the solution must act together, and the removal of either one will result in the arrest of propagating cracks and the prevention of crack initiation.
3.   Cracking can be either intergranular or transgranular.

Many of the alloys observed to suffer from SCC are normally covered with a highly protective oxide film or passive layer. Once this film has been ruptured, the plastic deformation produced at the tip of the stress corrosion crack prevents complete repassivation, thus allowing metal-environment reactions to occur. The rate of repassivation is an important factor, particularly with titanium alloys in aqueous chloride solutions. Initial tests showed that these alloys were immune to SCC, but later experimentation showed that if a sufficiently high strain rate could be maintained in the crack tip region, some degree of susceptibility resulted. The prinicpal mechanisms proposed for SCC are

1.   Accelerated dissolution of yielding metal at the crack tip
2.   The formation and subsequent fracture of a brittle oxide film
3.   Hydrogen embrittlement due either to the formation of hydrides or to the trapping of dislocations by hydrogen ions, leading to a loss of ductility
4.   Stress sorption mechanisms which rely on the lowering of bond energies in the crack tip region as a result of the absorption of a critical species from the solution
5.   Selective attack at grain boundaries due possibly to impurity segregation or precipitation, the function of the stress being to pull the weakened material apart

In 18% Cr-8% Ni stainless steel, failure can be either transgranular or intergranular, the latter frequently being associated with sensitized material. Transgranular cracks can, however, be initiated by intergranular cracks. Elevated temperatures favor transgranular cracking in stainless steel so that care must be taken during sterilization procedures to avoid critical combinations of stress, temperature, and environment.

In titanium alloys, failure is transgranular and is due either to hydrogen embrittlement or a dissolution mechanism. Although chromium cobalt alloys are most unlikely to suffer from SCC in the body, premature failure can result due to the chemical sharpening of notches, which may already be present in the design of the component, cast chromium cobalt alloys in particular having very low resistance to notch impact failure.

B. Corrosion Fatigue

A material subjected to repeated stress cycles may be susceptible to fatigue cracking, depending on the amplitude and number of the stress cycles. If the material is also exposed to a corrosive environment, then the fatigue cracking may be accelerated and the process termed corrosion fatigue. The effect is illustrated in Figure 11 where a graph of the applied stress vs. the number of applied local cycles to failure (an S-N curve) is shown for materials in aggressive and nonagressive environments.

Good fatigue resistance is of paramount importance, particularly for implants in the lower limbs, especially the hip joint. As fatigue cracks always start at the surface of a component, the surface condition is most important. Scratches, notches, or pits may lead to fatigue crack initiation and subsequent component failure. Placing the surface under a compressive stress, for example, by shot peening will frequently improve the fatigue life.

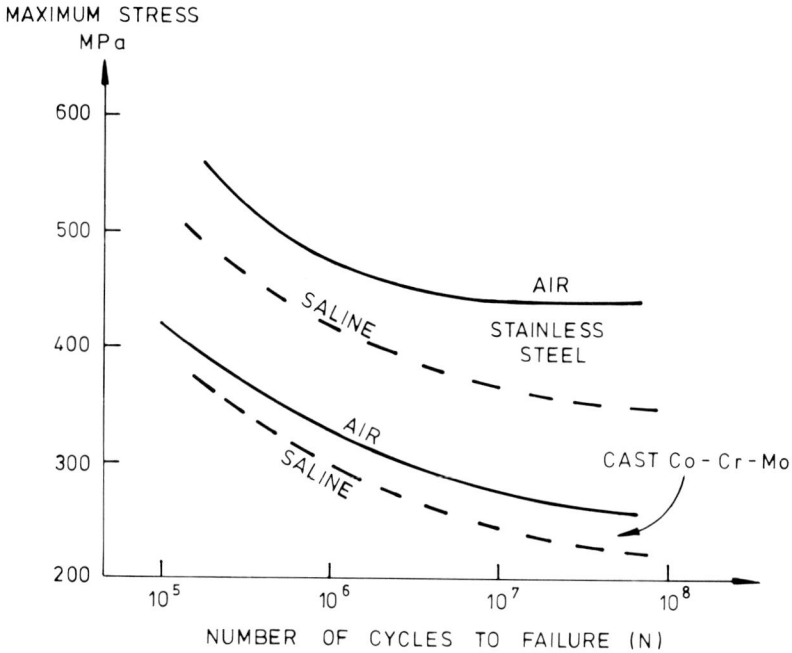

FIGURE 11. The effect of a saline environment upon the fatigue resistance of stainless steel (cold worked) and a cast Co-Cr-Mo alloy.

As with SCC, corrosion fatigue is a complex interaction between the metal and the stress, with the rupture of protective oxide films again playing an important role. A close link exists between corrosion fatigue and SCC, and many of the mechanisms proposed to account for the latter have been adapted to encompass failure under fluctuating loads.

C. Fretting Corrosion

Fretting corrosion is the combination of surface degradation occurring due to relative oscillatory tangential slip of small amplitude between components and the action of a corrosive environment. The relative amplitude is low (<75 μm) and the relative velocities are much lower than in wear, giving rise to the trapping of fretting debris between the surfaces.

Fretting corrosion is most probably a combination of abrasive and corrosive wear acting conjointly, and abrasion, caused by the entrapment of the worn off oxide film and oxidized wear particles between the surfaces, ruptures the oxide film, thus allowing further corrosion to occur.

Adhesive wear is one of the most prevalent types of wear and is the cold welding of minute surface asperities due to plastic flow resulting from the high stress encountered by these small areas of contact. Abrasive wear occurs when a hard body ploughs or gouges a softer one. In fretting, the harder surface may be one of the components or a third body, for example, an oxidized wear particle. In general, the harder the material, the greater is the resistance to abrasive wear. Corrosive wear differs from abrasive wear in that corrosion occurs at the bare metal surface after the removal of the protective film mechanically.

In stainless steel implants, fretting corrosion produces scars with a pitted or granular appearance as distinct from the more polished wear traces. Angular corrosion pits are found at the periphery of the fretting scars. Deep corrosion tunnel-pits are observed in some scars, and chemical attack occurs along selected crystallographic planes leading to etch pitting.

By contrast, chromium-cobalt-molybdenum alloys show evidence of polished scars, corrosive activity varying from intense local attack to more general corrosion. The scars possess a rippled appearance possibly caused by grooves worn in the material by an abrasive wear process. Some grooves may contain fine rounded particles, possibly the result of a breakup of brittle carbide particles exposed to the fretting surface.

Titanium alloys possess poor wear properties and finely divided metallic titanium is believed to be a product of the wear process. Fretting scars have a plowed appearance rather than polished or pitted, and cratering may also be observed.

VI. CORROSION IN PRACTICE

In the following section, the more practical aspects of the corrosion of implants will be considered. Of the three groups of metals which are in current use as surgical implants, both theory and practice have indicated that corrosion resistance increases with the order stainless steel, cobalt-chromium alloys, and titanium-based alloys. This was shown experimentally in a study of the variation of isolated material potential with time for samples exposed to aqueous saline solutions. The potential of stainless steel specimens varied with time in a manner indicative of continued film breakdown and subsequent repair, indicating that under service conditions the alloy would be unable to resist breakdown by pitting. With both cobalt-chromium alloys and titanium-based alloys, the isolated potential remained stable with time. It was concluded that pitting attack would be resisted for very long periods for cobalt-chromium alloys and for an indefinite period for some titanium-based alloys.

Clinical experience has shown that for titanium and titanium-based alloys, corrosion presents few problems. Blackening of the tissues surrounding some titanium implants has been observed, but this is believed to be the result of wear processes rather than corrosion. Increased levels of titanium in the surrounding tissue have also been observed, although in neither phenomena have the effects been clinically significant. Titanium owes its excellent corrosion resistance to a very stable protective oxide film which if broken reforms very rapidly in aqueous solutions, preventing corrosive attack. The combination of a stress in the region of a sharp notch or defect or continued exposure to fluctuating loads can lead to mechanical rupture of the protective film. Under these conditions titanium alloys may be susceptible to SCC or corrosion fatigue. Very few examples of corrosion of cobalt-chromium alloys have been reported in the literature, indicative of the good corrosion resistance as predicted by experiment.

In contrast, there is much evidence in the literature of examples of corrosion of stainless steel implants, the attack taking many of the different forms described previously. Of necessity therefore this section will be almost exclusively confined to examples of the corrosion of stainless steel implants. The following features are frequently contributory factors in the initiation or acceleration of the corrosion process:

1. Incorrect metallurgical composition
2. Incorrect metallurgical condition
3. Poor design or incorrect useage of implants
4. Surface finish of implants

A. Effects of Metallurgical Variables

As type AISI 316L, the principal austenitic stainless steel used for surgical implant applications, is susceptible to corrosion in aqueous environments, it is of paramount importance that the chemical composition is maintained within the specifications laid down by the various standards organizations. The amounts of chromium, molybdenum, carbon, and nickel present in the steel are all important if the material is to have adequate corrosion resistance.

The presence of molybdenum within the limits of 2.0 to 4.0% is of the greatest significance.

Too high a concentration may lead to the formation of a σ phase, which can result in embrittlement of the steel, while too little enhances the possibility of corrosion, particularly pitting attack. In view of the importance of this, it is both surprising and disappointing to note that there is a large percentage of implants removed due to excessive corrosion which were subsequently found to be deficient in molybdenum. In some implants, no molybdenum was present, possibly indicating that an incorrect grade of steel (for example, AISI 304) had been used and not a faulty batch of the correct material.

Variations in the nickel content affect both mechanical properties and the corrosion resistance. Reducing the concentration of nickel increases the possibility of galvanic corrosion in multicomponent devices. As discussed earlier, this phenomenon is not restricted to different materials, but may occur when components, made from different compositions or with different microstructures, are in electrical contact. This frequently occurs in bone plates and screws, where manufacturing techniques involving cold deformation can lead to martensitic transformations in screws which may then corrode preferentially when in contact with the austenitic plate. This process would be reduced if the nickel content were kept above 11% as this retards the martensitic transformation. High nickel contents, however, may result in screws of inferior mechanical properties due to the associated decrease in work hardening rate. These two effects of nickel concentration may imply that the range of concentrations allowed should be reduced. Concern for the corrosion properties of the steel has been shown by the increase in recent years in the minimum concentration of nickel allowed in the steel by BSI from 8 to 10%, although this may still be too low as 11% is felt by some workers to be a critical concentration. Very few cases of steel depleted in chromium have been reported possibly because this element is standard to all stainless steels.

B. Incorrect Metallurgical Condition

When considering incorrect metallurgical condition, sensitization, which has been discussed previously, is of paramount importance. A continuous network of grain boundary precipitates provides a path for corrosion and a means of deep penetration of the implant. Removal of whole grains may occur, resulting in serious weakening or fracture. Figure 12 shows the sensitized structure of a failed stainless steel osteotomy plate removed after 3 years in a patient. Intense grain boundary attack had occurred as shown in Figure 13 which also shows grain boundary fall out at the specimen surface. Although the techniques for avoiding the sensitization of stainless steel have been well documented for many years, it is only recently that the problem has been successfully overcome. The most recent trend of producing steels of low carbon content has also helped to improve the situation.

Although of great significance, sensitization is not the only undesirable metallurgical condition which may be observed in stainless steel implants. In welded components, delta ferrite may be present which although not considered harmful for many applications can lead to increased corrosion rates and is both unnecessary and undesirable in implants. The presence of nonmetallic inclusions, particularly sulfides, may also reduce the resistance to corrosive attack. Figure 14 shows the large globular inclusions observed in a pitted and cracked gluteal post from an EN 58J osteotomy plate which also had a microstructure most unusual in that type of steel. Special melting techniques which are being used with increasing frequency will help to reduce this problem as they enable cleaner steels to be manufactured with significantly lower inclusion contents.

C. Poor Design and Use of Implants

Poor design or the incorrect use of implants is principally apparent in multicomponent devices, leading to the formation of crevices and notches which then promote corrosive attack. Bone and nail plates present particular problems in this area due to the multiplicity of design combinations available for plates and screws. The effects of different designs of

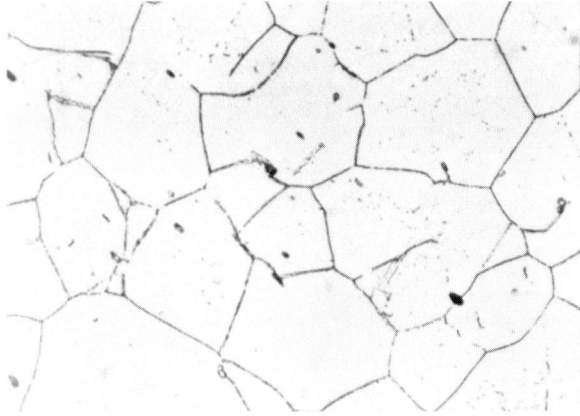

FIGURE 12.   Sensitized stainless steel microstructure observed in an osteotomy plate. Grain boundary carbides are clearly visible. (Magnification × 385.)

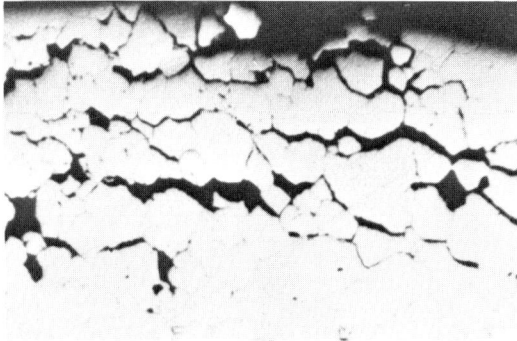

FIGURE 13.   Intense grain boundary attack and grain fallout occurring in the sensitized plate shown in Figure 12. (Magnification × 100.)

screw heads with respect to corrosion performance have been studied. Three aspects of screw design were examined from the viewpoint of minimizing corrosion attack: head design, underhead design, and the finish of the threads.

The principal effect of head design upon the corrosion behavior is the likelihood of damage caused by the screwdriver slipping out of the head (cam-out) during insertion. This may then scratch the implant and lead to a possible disruption of the oxide film. Metal transfer may also occur between the screwdriver and the implant, leading to the possibilities of galvanic attack, although this is now less likely with the increased use of screwdriver bits made from the same material as the implant. The principal types of head design are single slot, cross slot, Phillips, uniform, Posidrive, and hexagonal recess, with the propensity for cam-out decreasing from single slot to hexagonal as the area of vertical driving faces increases.

The three main types of underhead design currently used are flat, conical countersink, and spherical countersink. Although all types of design give excellent alignment (and consequently limited access for crevice corrosion) when there is near perfect alignment with the pilot hole, problems arise due to misalignment and ovality of the screw threads. With

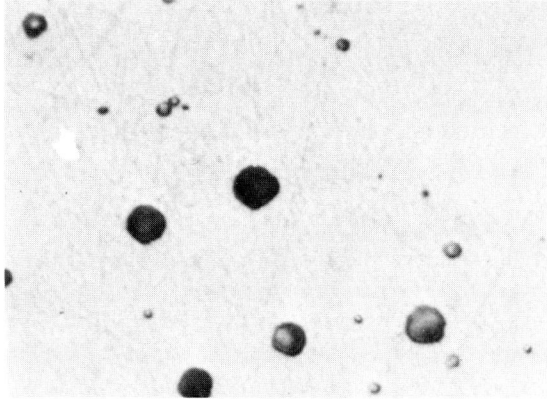

FIGURE 14. Large globular inclusions observed in a gluteal post (EN58J) of an osteotomy plate. (Magnification × 1150.)

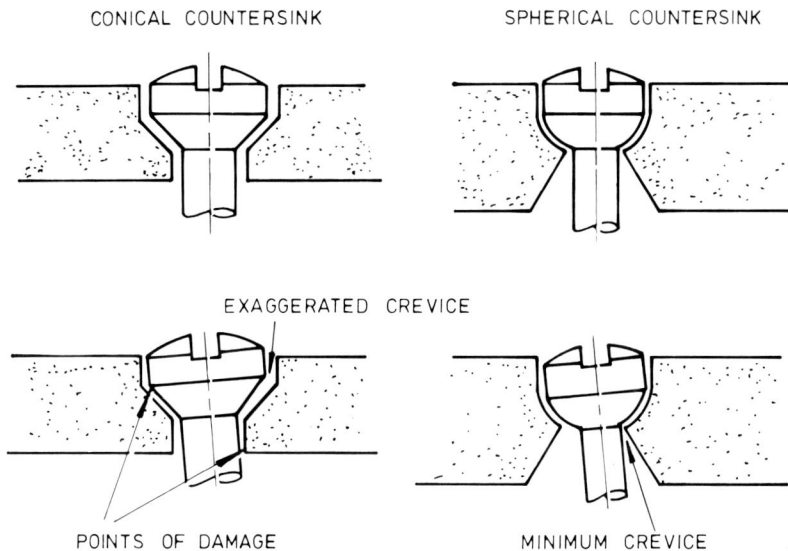

FIGURE 15. Example of crevice formation with the use of poorly aligned screws. This is minimized by using spherical countersunk screws.

flat or conically countersunk underhead designs, a small degree of misalignment leads to the seating being limited to small areas which are then subjected to high pressures which may in turn lead to surface damage and possible fretting corrosion.

The use of a spherical countersink, in conjunction with a spheroconical plate hole, allows 15 to 20° misalignment of the screw while still giving good seating at the countersink. This results in even pressures at the countersink and reduces the propensity for corrosion. Figure 15 shows the effects of misalignment upon the probability of crevice formation when using spherical or conical countersinks. The use of the former clearly allows a much higher degree of misalignment before a crevice is formed.

Although the design of the threadform would appear to have little effect upon corrosion behavior, the surface finish is of paramount importance, with electropolishing being necessary, especially if a thread rolling technique has been used, which may lead to a highly work hardened surface containing inclusions, folds, and rokes.

Many design problems relate to interfaces which can result in crevice formation, pitting attack, or fretting corrosion. Components which have many points of contact frequently exhibit evidence of corrosive attack as was noted, for example, in the spinal plate shown in Figure 16. Although the pitting did not contribute to the mechanical failure of the device, had the plate remained intact, further problems might have resulted from interactions between the corrosion product and the surrounding tissue. Reducing the amount of corrosion could be achieved either by redesigning the component, minimizing the number of points of contact, or by manufacturing it from a material less susceptible to pitting corrosion.

It has been suggested that the use of inert spacers between the components would reduce the propensity for interface corrosion. Such a move might, however, lead to increased attack, as was noted with gluteal posts for osteotomy plates, due to the difficulty of eliminating entirely the ingress of solution between the component and the spacer. Aspects of design which lead to stress concentrations thereby increasing the probability of failure due to corrosion-stress interactions are considered in the following section.

D. Effects of Surface Finish

There are two aspects of surface finish which are of importance to corrosion studies. These are the effects of surface treatments or blemishes upon the propensity for corrosive attack and the probability of initiating corrosion-stress interrelated phenomena.

One of the most important surface treatments applied to stainless steel implants is passivation in nitric acid which has long been recognized as giving improved corrosion treatment. The process which is also applied to titanium alloys and which has the additional advantage of removing any iron particles which may have become embedded in the implant surface during polishing consists of immersion in a 30% nitric acid solution at 40 to 60°C for approximately 15 min. More recently, it has been proposed that passivating in oxygenated isotonic saline solutions can give enhanced corrosion resistance.

Polishing either by mechanical or electrochemical methods is frequently used to improve the surface finish. The latter gives a superior corrosion resistant surface, possibly due to the method and nature of the treatment. Mechanical polishing leaves a slightly work hardened layer on the surface, while electropolishing (which may be regarded as a controlled electrochemical corrosion process) preferentially removes high energy atoms, for example, kink sites or ledges, which might be expected to undergo accelerated attack.

After polishing, the effects of isolated scratches incurred, for example, during handling or insertion, depend significantly upon their position on the implant and the ability of the damaged area to repassivate. Where repassivation is possible, only a short-term increase in the corrosion rate would be expected. If, however, the damage were in a crevice a permanent site of attack might result. A further factor to be considered is how the scratch was formed; contact with a harder material, for example, leads to the possibility of fretting corrosion, metal transfer, and galvanic attack.

The surface finish is also of importance for both titanium and chromium-cobalt alloys, neither of which, as noted earlier, experience serious corrosion problems. The blackening of the tissue observed around some titanium implants may be overcome by anodizing (growing a thick oxide film on the surface) which improves the wear resistance. One of the few examples of corrosion in cobalt-chromium alloys was crevice attack associated with large amounts of porosity at the surface. Such porosity is the result of shrinkage or poor feeding techniques during casting and is observed on occasions in cast implants.

Surface finish is also of considerable importance when considering resistance to corrosion fatigue or SCC. Fatigue crack initiation is a surface phenomenon and usually occurs in a region of maximum stress. Once a crack has initiated it will, in most circumstances, continue to propagate until complete failure of the device occurs. Any surface defect, for example, polishing or grinding marks, may be sufficient to promote cracking if it occurs in a region

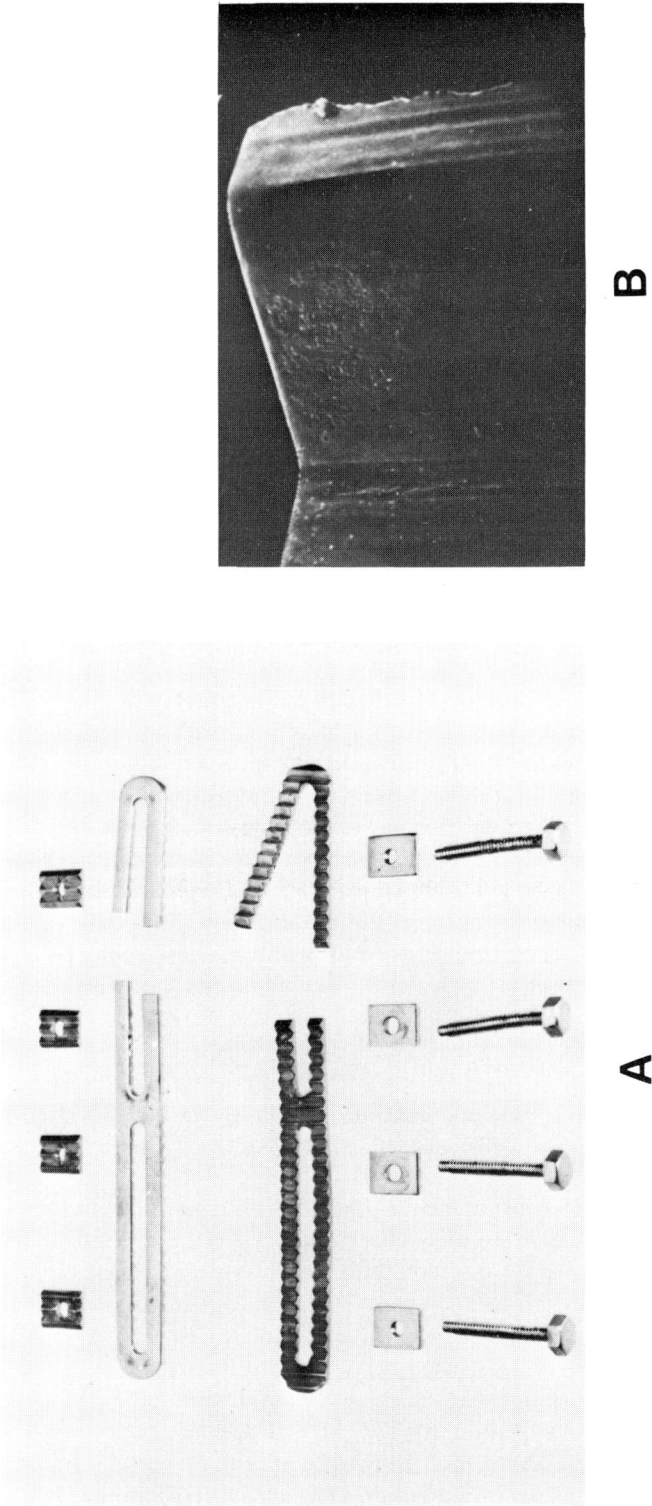

FIGURE 16. (A) Spinal brace of stainless steel showing many points of contact and many areas for crevice attack; (B) pitting corrosion of the plate at points of contact between plate and washer.

of high stresses. The effect of the defect may be to act as a precrack or stress concentrator, increasing the stress intensity at that point. Examples have been noted of surface defects which have initiated fatigue cracks, notably in stems of artificial hip joint prostheses.

Lack of appreciation of the above has led to occasional examples of cracks starting at defects deliberately introduced into the implant in highly stressed regions. Two cases have been reported of failed artificial hip joint prostheses which have broken on the stem due to cracks initiated at spark-etched trade marks on the lateral surface of the stem in the region of the maximum stress. Marks induced by this method produce very sharp notches which consequently have large stress-raising effects. These examples together with other design features, sharply notched serated edges, sharp corners, and sharply radiused edges, may all serve as suitable stress concentrations to damage the oxide film and so help initiate corrosion fatigue or SCC.

The incidence of "failures" is low, but what emerges from many of the examples of components that have failed due to excessive corrosion is that in many cases the factor or combination of factors which contributed to the attack could have been avoided. Failure to produce steel to the required compositional standards or in the required metallurgical condition implies insufficient control over the manufacturing processes and ineffective quality control which allows such implants to be passed as satisfactory. Quality control can also be questioned when surface defects in highly stressed regions remain undetected. Improvements in the quality of implants is being achieved as realistic standards are being produced and as a result of the cooperation and understanding being achieved between surgeons and materials engineers.

Marking of implants in highly stressed regions and in such a manner as to produce sharp notches indicates a deep-rooted lack of understanding of the problems involved which can only be overcome by increased awareness by all involved with surgical implants of the principles governing the mechanical properties and corrosion behavior of the product when in service. Where failure is the result of poor component design or incorrect material selection, the remedy lies with product ion and improvement or use of a different alloy. In this area, new designs are continually being adopted frequently using computer aided designs. A detailed program of standardized testing may be required to prevent the large-scale production of implants containing fundamental design faults or which utilize unsatisfactory material. The frequency of failed devices due to excessive corrosion can be reduced, but only by greater vigilance and effort by all associated with surgical implants.

VII. OTHER METALLIC MATERIALS

In addition to 316L stainless steel, titanium and titanium alloys, and the wrought and cast chromium-cobalt based alloys so far considered, other metals have lesser although important uses in orthopedic surgery. The most important of these are other high-strength stainless steels, often employing cold work to increase their yield strength, other chromium-based alloys, and tantalum. The corrosion behavior of these metals will therefore be briefly examined.

As the corrosion and pitting resistance of 316L stainless steel is, under certain circumstances, barely adequate, any effects of compositional variation or work hardening upon the corrosion behavior are most important. Higher molybdenum content steel (AISI 317) has been claimed to offer increased resistance to crevice corrosion, but its resistance to pitting attack has been found to be no greater than that of conventional AISI 316L. Increases in proof stress of up to 25% in comparison to that of AISI 316L can be achieved by the addition of 0.2% nitrogen (AISI 316N) without any loss of corrosion properties. Similar increases in yield strength can also be achieved by prior deformation (cold work), but only at the expense of the already marginal corrosion resistance. The electrode potentials at which crevice corrosion and pitting attack can occur are lowered by up to 250 mV after 30% cold work, due presumably to the presence of increased numbers of dislocations and deformation bands.

This limitation upon the capacity to increase the strength of steels based upon the 18% Cr-10% Ni family has led to other steels being examined as potential orthopedic materials. Duplex steels based on 25% Cr and 5% Ni with a microstructure of 50% austenite and 50% ferrite possess good wear resistance and have superior mechanical properties when compared with the 18-10 family. Both wrought and cast types are available. Although the pitting potential of these alloys is generally higher than the 316 type material, repassivation potentials (the potential required to repassivate an active pit or crevice) are comparatively low and the passive current densities are high, increasing the likelihood of patient sensitivity to the implant. The reproducibility of results for this material is poor (in comparison to stainless steels in general which are characteristic in their poor reproducibility of corrosion data), due presumably to the compositional variations and the inclusions present in the commercial grades of the material. This poor reproducibility gives rise to occasional very low values of pitting and repassivation potentials, indicating that particular implants might experience excessive corrosion. If such problems can be overcome by stricter compositional control, the material does appear to offer an attractive alternative to the conventional austenitic stainless steels.

High-ductility transformation-induced plasticity (TRIP) stainless steels based upon 7% Ni, 12% Cr, and 3 to 4% Mo have also been examined as potential orthopedic materials as these would permit implant size reduction and design refinements. Although such materials have shown promising results in laboratory trials, with adequate resistance to corrosion fatigue, stress corrosion, and pitting attack, clinical trials have indicated inferior resistance to pitting attack and general corrosion in comparison to cold-worked AISI 316L. Evidence of intergranular stress-corrosion cracking has also been observed. The corrosion resistance of such steels is, however, very dependent upon the molybdenum content of the steel and upon the level of cold work, and other combinations of these variables may produce materials which can have potential as orthopedic alloys.

Such higher strength steels do therefore appear to offer possible alternatives to the 18% Cr-10% Ni family, but manufacturing practice must be closely controlled to avoid excessively high impurity contents and variations in composition. In addition, the compositions which offer the maximum corrosion resistance must be fully established prior to in vivo testing.

Although the cast Co-Cr-Mo and wrought Co-Cr-Ni-W alloys considered previously both possess good mechanical properties and excellent general corrosion resistance, other alloys based upon the Co-Cr system are used in orthopedic surgery. An example of these is MP-35N (20% Ni, 10% Mo, 20% Cr, 50% Co) which after solution treatment has an austenitic matrix, but undergoes transformation to ϵ-martensite during cold working. Precipitation heat treatment can also produce an extrahard version of the material. The yield strength is typically in excess of 500 MPa (compared with 200 MPa for AISI 316L and 300 MPa for wrought Co-Cr-Ni-W alloys), and the wear resistance is sufficient for it to be used as the femoral head in artificial hip joint prostheses. The general corrosion resistance of the material is extremely high, and excellent resistance to stress corrosion cracking and corrosion fatigue has been reported. In addition the pitting, crevice and depassivation resistance is claimed to be higher than for AISI 316L and the cast cobalt-based alloys. Electrochemical corrosion studies, however, have shown that the cast Co-Cr-Mo alloys may possess higher pitting potentials than either the wrought Co-Cr-Ni-W alloys or MP-35N. Indeed while stable passivity is predicted for the wrought and cast alloys, long-term potential time measurements made on MP-35N have indicated that some local breakdown of passivity can occur, although a general breakdown would not be expected under normal circumstances. While the onset of localized pitting attack might therefore be expected in MP-35N under crevice or similar conditions, this would not be expected to result in a total breakdown in passivity.

Although the resistance of MP-35N to crevice corrosion is greater than that of 316L (the most susceptible of the principal orthopedic metals), there are indications therefore that it is less than that recorded for the more commonly used cast and wrought cobalt based alloys. MP-35N can, however, be used safely in double alloy prostheses (coupled with, for example, cast Co-Cr-Mo or titanium or titanium alloys) without the likelihood of galvanic attack in either metal.

There are several high-strength wrought alloys based on the Co-Ni-Cr-Mo system, all of which exhibit the excellent general corrosion resistance associated with these alloys. Under the more stringent conditions of crevices, however, there remains the possibility of localized breakdown of passivity (with the attendant problems of loosening of the implants and patient sensitivity to the metal ions resulting from the corrosion process), which appears to be increased with the higher strength materials.

Tantalum has long been used in orthopedic surgery in the pure, unalloyed state for nonload-bearing applications principally as clips, wires, and foil in neurosurgery. Very few data are available, however, concerning its general corrosion resistance and more specifically the resistance to pitting and crevice attack. The lack of tissue response to the solid material does, however, indicate that tantalum possesses inherently good corrosion resistance. Its tendency to fracture and disintegrate when used in gauze form and its lack of mechanical strength have contributed to a decline in its popularity and usage in recent years.

VIII. DEGRADATION OF NONMETALLIC MATERIALS

The preceding sections have considered the corrosion of metallic surgical implants; any review of the effects of the body on implanted materials would, however, be incomplete without some reference to the wide variety of nonmetallic materials used in prosthetic devices. These include alumina ceramics, polymethylmethacrylate, ultrahigh molecular weight polyethylene, nylon, and bioglasses.

Much less is known about the effects of the physiological environment upon nonmetallic materials than for metals. This is partly the result of the concentration of scientific effort over recent years into a limited number of metallic materials suitable for surgical applications, characterizing their behavior in some detail. In contrast, an increasing number of polymeric and other nonmetallic materials are being introduced, with the result that knowledge concerning each particular group increases only slowly. In addition there is greater difficulty in standardizing polymeric materials so that practical and laboratory experience gained in different locations is rarely on precisely the same material. It is perhaps not surprising therefore that the effects of the physiological environment upon polymeric materials are much less well documented than is the case for metals, and relatively few experimental studies have been performed.

Nonmetallic materials do not suffer from electrochemical corrosion, but may experience a slow change in properties with increasing time of exposure to physiological solutions. This change is often a surface phenomenon and is referred to as degradation. In the following section, a brief account of the effects and cause of degradation will be presented for polymeric materials, ceramics, and bioglasses.

A. Polymeric Materials

The effects of physiological media upon polymers are usually evident, either as a loss of mechanical properties, notably tensile strength which may be reduced by as much as 80%. The five principal mechanisms of general polymer degradation are

1. **Random chain scission** — This is the fragmentation of the molecular chains, leaving fragments which are large in comparison with the monomer unit. This leads to a rapid decrease in the molecular weight and a reduction in strength.

2. **Depolymerization** — This is similar to (1) above, involving the breaking of bonds on the basic carbon chain, but it is attacked at weak links in the chain or at the chain ends. The attack propagates along the chain until it is reconverted to a monomer or a secondary process stops the breakdown. An example of this is the thermal degradation of polymethylmethacrylate. During the polymerization of methylmethacrylate, the chain terminates by disproportionation of radicals. Two chains, each with unsaturated bonds ($\cdot$C) may react, terminating one chain by the addition of a hydrogen atom and the other by the loss of a hydrogen atom with the subsequent formation of a double bond.

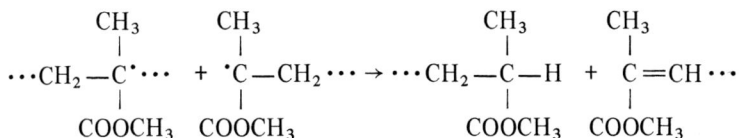

$$\cdots CH_2 - \overset{\overset{\textstyle CH_3}{|}}{\underset{\underset{\textstyle COOCH_3}{|}}{C}}{\cdot} \cdots + {\cdot}\overset{\overset{\textstyle CH_3}{|}}{\underset{\underset{\textstyle COOCH_3}{|}}{C}} - CH_2 \cdots \rightarrow \cdots CH_2 - \overset{\overset{\textstyle CH_3}{|}}{\underset{\underset{\textstyle COOCH_3}{|}}{C}} - H + \overset{\overset{\textstyle CH_3}{|}}{\underset{\underset{\textstyle COOCH_3}{|}}{C}} = CH \cdots$$

Chain ends of this type can suffer depolymerization as the double bond at the end of the chain is relatively weak and susceptible to environmental attack, with the production of the monomer.

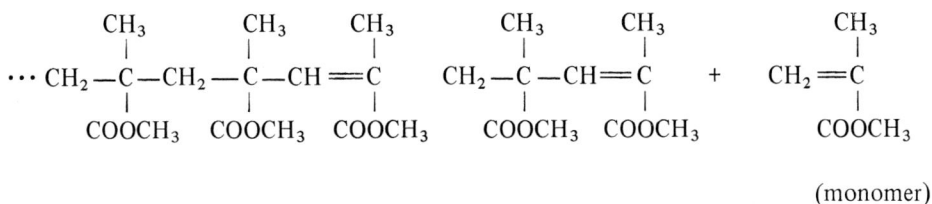

$$\cdots CH_2 - \overset{\overset{\textstyle CH_3}{|}}{\underset{\underset{\textstyle COOCH_3}{|}}{C}} - CH_2 - \overset{\overset{\textstyle CH_3}{|}}{\underset{\underset{\textstyle COOCH_3}{|}}{C}} - CH = \overset{\overset{\textstyle CH_3}{|}}{C} \quad CH_2 - \overset{\overset{\textstyle CH_3}{|}}{\underset{\underset{\textstyle COOCH_3}{|}}{C}} - CH = \overset{\overset{\textstyle CH_3}{|}}{\underset{\underset{\textstyle COOCH_3}{|}}{C}} + CH_2 = \overset{\overset{\textstyle CH_3}{|}}{\underset{\underset{\textstyle COOCH_3}{|}}{C}}$$

(monomer)

The process becomes autocatalytic, as the new chain also terminates in an unsaturated radical; chain irregularities may, however, terminate the process.

3. **Cross-linking** — This process is the formation of bonds across molecular chains, and the most serious result is the subsequent embrittlement of a normally rubbery polymer. It is most likely to occur during sterilization by ionizing radiation and may also result in component shrinkage. One of the best-known examples of this is the effect of radiation on the polyethylene chain, which results in the ionization of the hydrogen atoms.

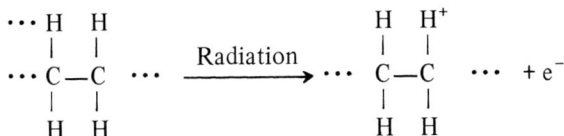

$$\cdots \overset{\overset{\textstyle \cdots H \quad H}{|\quad\ |}}{\underset{\underset{\textstyle H \quad H}{|\quad\ |}}{C - C}} \cdots \xrightarrow{\text{Radiation}} \cdots \overset{\overset{\textstyle H \quad H^+}{|\quad\ |}}{\underset{\underset{\textstyle H \quad H}{|\quad\ |}}{C - C}} \cdots + e^-$$

If the electron can escape, liberation of the hydrogen atom and the formation of a free radical can occur.

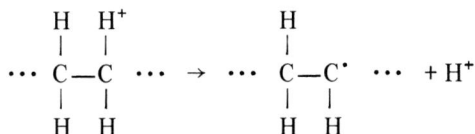

$$\cdots \overset{\overset{\textstyle H \quad H^+}{|\quad\ |}}{\underset{\underset{\textstyle H \quad H}{|\quad\ |}}{C - C}} \cdots \rightarrow \cdots \overset{\overset{\textstyle H}{|}}{\underset{\underset{\textstyle H \quad H}{|\quad\ |}}{C - C{\cdot}}} \cdots + H^+$$

Cross-linking can then occur between the radicals.

$$
\begin{array}{ccc}
H & H & H \\
| & | & | \\
C-C-C^{\bullet} \\
| & | \\
H & H
\end{array}
\qquad
\begin{array}{ccc}
H & H & H \\
| & | & | \\
C-C-C \\
| & | \\
H & H
\end{array}
$$

$$
+ \qquad \rightarrow
$$

$$
\begin{array}{ccc}
H & H \\
| & | \\
C-C-C^{\bullet} \\
| & | & | \\
H & H & H
\end{array}
\qquad
\begin{array}{ccc}
H & H \\
| & | \\
C-C-C \\
| & | & | \\
H & H & H
\end{array}
$$

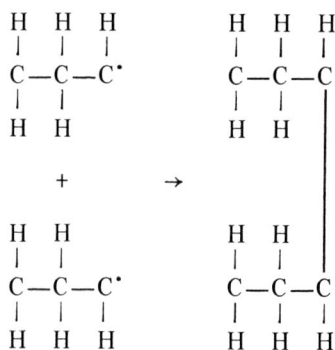

4. **Bond changes** — This involves changes in the nature of the bonds between carbon atoms on this basic molecular chain, without cross-linking or chain scission. An example of this is the formation of double carbon-carbon bonds from single bonds, a process which would be progressive in the thermal degradation of PVC, due to a lowering of the bond strength of the hydrogen and chlorine atoms attached to adjacent carbon atoms.

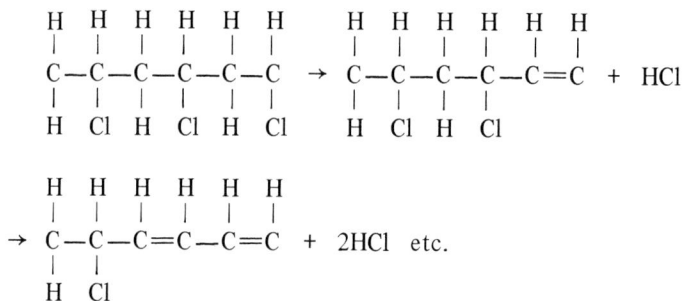

$$
\begin{array}{cccccc}
H & H & H & H & H & H \\
| & | & | & | & | & | \\
C-C-C-C-C-C \\
| & | & | & | & | & | \\
H & Cl & H & Cl & H & Cl
\end{array}
\rightarrow
\begin{array}{cccccc}
H & H & H & H & H & H \\
| & | & | & | & | & | \\
C-C-C-C-C=C \\
| & | & | & | \\
H & Cl & H & Cl
\end{array}
+ \ HCl
$$

$$
\rightarrow
\begin{array}{cccccc}
H & H & H & H & H & H \\
| & | & | & | & | & | \\
C-C-C=C-C=C \\
| & | \\
H & Cl
\end{array}
+ \ 2HCl \ \ \text{etc.}
$$

5. **Side group changes** — These can occur without significant changes in the molecular weight of the material and frequently involves hydrolysis. An example of this is the conversion of poly(vinyl acetate) which is insoluble in water to poly(vinyl alcohol) which is water soluble, as a result of the catalytic action of water.

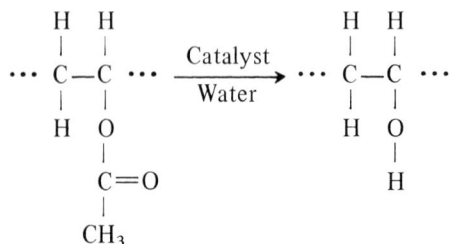

$$
\cdots
\begin{array}{cc}
H & H \\
| & | \\
C-C \\
| & | \\
H & O \\
& | \\
& C=O \\
& | \\
& CH_3
\end{array}
\cdots
\xrightarrow[\text{Water}]{\text{Catalyst}}
\cdots
\begin{array}{cc}
H & H \\
| & | \\
C-C \\
| & | \\
H & O \\
& | \\
& H
\end{array}
\cdots
$$

The principal degradation agencies are heat, light, and oxygen, none of which would be expected to be active within the body, but may be operative during storage or sterilization procedures. Reactions with water (hydrolysis) have been considered as the most important with respect to in vivo polymer degradation. The polymers most susceptible to degradation in the body therefore are hydrophilic materials which contain hydrolyzable linkages. Examples of these are some nylons, for example, polyhexamethyleneadipamide.

A further aspect of polymer degradation is the deterioration resulting from the specific activity of living organisms. Although no cases have been observed in implanted polymers, several well-documented areas of biodeterioration indicate that this is a real possibility. Degradation is not the only reaction of significance to occur; the compatibility of the potentially leachable additives within the plastics are also of importance and plasticizers of low toxicity must therefore be used.

As with metals, the combination of a stress and an environment can produce accelerated failure. One of the most serious forms is environmental stress cracking, which is prevalent in low molecular weight polyethylene. The stress and environment combine to produce brittle cracking with no visible effects of chemical interaction, for example, swelling.

A further stress-environment interaction is stress solvent crazing where solvents have an accelerated effect upon the plastics in the presence of a stress. Polymethylmethacrylate in particular suffers from this effect which has been observed in acrylic Judet replacement hip prostheses.

Sterilization techniques may also lead to a deterioration in mechanical properties. Dry heat sterilization is unsuitable for most plastics as softening or deformation may occur. Steam sterilization involves lower temperatures, although creep and deformation may still result. Gamma radiation is frequently used even though all polymers are affected to some extent. Both random chain scission or cross-linking are promoted by radiation. Gaseous sterilization using, for example, ethylene oxide is a further alternative, although some reactions with functional side groups may occur.

B. Ceramics

It is only recently that ceramics have been used in surgery and consequently little experimental data are available assessing their performance in the human body. One of the potential advantages of ceramics is that they can be relatively easily prepared with a porous surface, thus allowing bonding by tissue ingrowth.

When exposed to physiological media or distilled water, ceramics suffer a loss of strength and a decrease in fatigue properties, the effects being related to the porosity of the ceramic. The activation energy for the process is stress dependent, with grain boundaries the most likely site for the reaction. The decrease in properties is believed to be related to the transfer of calcium and silicon ions from the ceramic to the solution which undergoes an increase in the concentration of these ions during testing.

Steam autoclaving induces a substantial loss of mechanical properties in ceramics with calcium segregation being observed at both external and fracture surfaces. The decrease in properties can be correlated with the extent of calcium segregation.

It has been proposed that porous ceramic coatings may be bonded onto the metal stems of artificial hip prostheses. This would allow bone ingrowth and securely anchor the prosthesis. Studies of combinations of metals and ceramics, joined either by plasma or flame spraying, have shown that exposure to Ringer's solution reduces the bond strength at the metal-ceramic interface which is attributed to corrosion pitting. Determination of the composition of the surface of the alumina ceramic before and after exposure indicates that there is a substantial decrease in the amount of Al_2O_3 after aging with a corresponding increase in the amount of $Al(OH)_3$. The concentration of the calcium ion in the solution again increased during exposure. Use of a stainless steel substrate led to the appearance of rust spots, indicating penetration of the ceramic and attack on the metal.

These observations have shown that ceramics are less stable in the body than was at first believed and that further work is necessary in this area to resolve the problems before significant progress can be made in the application of ceramics to surgical requirements.

C. Bioglasses

Bioglass materials were developed to assist the bonding of tissues and biomaterials and are glass-ceramics with controlled surface ion activities which allow chemical bonding to the bone. Their composition and structure are of paramount importance as this controls the rate of release of products from the surface. Too low a release rate would not allow bonding to occur, while too high a rate might lead to cell necrosis.

Bioglasses are susceptible to both general corrosion and pitting attack in aqueous solutions, as they contain a silicate glassy phase. The corrosion process involves a proton from the water penetrating the glassy network and replacing an alkali ion which is then released to the solution. An OH^- ion is produced which destroys the Si-O-Si bonds, forming Si-O bonds which then interact with water. At pH values of less than 9, alkali ion release is more rapid and a silica-rich film develops on the surface which may go into the solution, resulting in continued attack. Stable SiO_2-rich films can form, however, resulting in protection from further attack. The release of silica is believed to be important in the initiation of bone mineralization and in producing the bond between the bone and the bioglass. As with ceramics, the use of bioglasses are at a relatively early stage and much work is required to develop more fully the theories explaining degradation in the body and to study in detail the effects upon the surrounding tissue.

GENERAL REFERENCES

1. **Evans, U. R.,** *The Corrosion and Oxidation of Metals,* Arnold, London, 1960.
2. **Pourbaix, M.,** *Lectures on Electrochemistry,* Plenum Press, New York, 1972.
3. **Scully, J. C.,** *The Fundamentals of Corrosion,* Pergamon Press, Oxford, 1975.
4. **Shrier, L. L.,** *Corrosion,* Vols. 1,2, Newnes, London, 1963.
5. **Colangelo, V. J. and Heiser, F. A.,** *Analysis of Metallurgical Failures,* Wiley — Interscience, New York, 1974.
6. **Syrett, B. C. and Acharya, A.,** *Corrosion and Degradation of Implant Materials,* ASTM STP 684, American Society for Testing and Materials, Philadelphia, 1979.
7. **Williams, D. F. and Roaf, R.,** *Implants in Surgery,* W. B. Saunders, London, 1973.
8. **Swanson, S. A. V. and Freedman, M. A. R.,** *The Scientific Basis of Joint Replacement,* Pitman Medical, 1977.
9. **Henthorne, M.,** Corrosion Causes and Control, Carpenter Technology Corp., Reading, Pa., 1972.
10. **Steigerwald, R. F.,** Electrochemistry of corrosion, *Corrosion* 1, 1, 1968.
11. **Hoar, T. P. and Mears, D. C.,** Corrosion resistant alloys in chloride solutions: materials for surgical implants, *Proc. R. Soc. London Ser. A,* 294, 486, 1966.
12. **Williams, D. F.,** Corrosion of implant materials, *Ann. Rev. Mater. Sci.,* 6, 237, 1976.
13. **Hughes, A. N. and Jordan, B. A.,** Metallurgical observations of some metallic surgical implants which failed in vivo, *J. Biomed. Mater. Res.,* 6, 33, 1972.
14. **Jordan, B. A. and Hughes, A. N.,** A review of the factors affecting the design specification and material selection of screws for use in orthopaedic surgery, *Eng. Med.,* 7, 114, 1978.
15. **Latimer, W. M.,** *Oxidation Potentials,* 2nd ed., Prentice-Hall, Englewood Cliffs, N.J., 1952.
16. **Makrides, A. C.,** *J. Electrochem. Soc.,* 104, 869, 1960.

Chapter 5

COATING AND SURFACE MODIFICATION

John H. Dumbleton and Paul Higham

TABLE OF CONTENTS

I. INTRODUCTION

Although much progress has been made in the development of new materials, there exists at present no single material which can withstand all the extreme operating conditions in modern technology. In general, components used in engineering applications must possess the appropriate structural characteristics to provide stiffness or flexibility and to carry applied loads without macroscopic failure. Such properties are associated with the bulk material of the component. However, the majority of engineering failures are surface initiated and occur through such phenomena as friction, wear, corrosion, fatigue, or a combination of these factors. That this should happen arises from two specific causes. In the first place, stress levels are often highest at the surface, and secondly, the surface is usually the only part which is subjected to the external environment. An increasingly common way of coping with such a situation is to provide different surface material properties from those of the bulk material. An everyday example concerns mild steel components electroplated with chromium to prevent corrosion. Surface treatment technology has advanced rapidly over the last few decades and has greatly assisted in meeting the ever increasing demands placed on materials.

Important surface treatment methods include coatings deposited by both chemical and physical methods. Such coatings are sufficiently adherent to the substrate to resist very severe stressing and are used as coatings for cutting tools. Electrodeposition, the oldest and most widely used coating technique, is still the subject of development and improvement. However, new techniques of producing modification of surfaces have assumed great importance. Examples of these new techniques are laser glazing and ion implantation. This review includes not only the established techniques, but also those techniques which gained prominence in the last few years.

In this review many techniques are covered along with the advantages and disadvantages of each. However, there are relatively few applications given in the orthopedic and other medical areas. In looking at the whole usage of materials for implants, and specifically at the orthopedics area, it must be realized that the process has been one of transfer and adoption from outside the medical and surgical areas. Materials developed for end-uses other than for implants have been utilized. The situation is particularly clear-cut for orthopedics in the cases of stainless steel (developed for general noncorrosion such as in chemical plants), titanium alloy (developed for aerospace applications), and cobalt-chromium-molybdenum alloy (developed for oxidation resistance at high temperatures). Even ultrahigh molecular weight polyethylene (UHMWPE) was developed for impact and abrasion resistance in mechanical applications and was only employed later, by chance, in total joint prostheses. It may thus be argued that a knowledge of the current state of the art in materials in general and in the coating/surface modification area in particular is needed so that the potential for transfer of these techniques to implant materials can be evaluated. The fact that examples of the use of these techniques can be cited for implant materials suggests that such a transfer is about to occur. There is thus the opportunity to shorten the usual time lag between material developments and the adoption of these developments in the implant areas.

The various methods by which a coating can be deposited onto a substrate are discussed in the following pages. This is followed by a discussion on the various applications of these treatments, which range from systems to prevent corrosion to those used in cardiovascular prostheses. Not all medical uses of surface coatings have been mentioned. Applications such as cementless fixation of orthopedic prostheses by the use of porous polymeric coatings are beyond the scope of this book, whereas the bioactive materials such as bioglass have been dealt with elsewhere.

II. TECHNOLOGY OF SURFACE TREATMENTS

A surface coating can be described as a covering applied to the outer face of an object, whereas surface modification is best described as changing the nature of the outer surface. For surface modification there is no sharp, or relatively sharp, interface between surface and bulk material. In the following, surface coating and surface modification will be treated separately.

A. Surface Coating Methods

In general, deposition processes may be divided into two basic types: (1) those involving droplet transfer, such as plasma spraying, arc spraying, and detonation gun coating, and (2) those involving an atom-by-atom transfer mode such as the physical vapor deposition or evaporation, sputtering, chemical vapor deposition, and electrodeposition.

1. Chemical Deposition

Probably the simplest example of chemical deposition is the ability of copper to plate out onto iron which is immersed in a solution of a copper salt. (This process has been utilized in copper mining areas where scrap iron is utilized to obtain copper from low concentration sources.) A second type of treatment is achieved autocatalytically, whereby the coating metal deposits on a metallic or metal-activated surface and the coating thickens with a more or less linear growth rate as long as the compositional balance of the solution is maintained. Solutions of this type are commonly known as ''electroless plating'' solutions.

Electroless coating processes are expensive to operate, but have the advantage that the coating thickness is uniform, irrespective of the complexity of the surface geometry. Adhesion of the coating is dependent upon chemical bonding, assisted by mechanical keying to a roughened surface.

Pretreatment prior to plating must be carefully carried out and varies with the substrate material. Steels should be electrolytically cleaned and acid etched to microroughen the surface. In the case of nonmetallic substrates (such as plastics), it is essential first to convert the surface of the nonmetal from the hydrophobic condition to the hydrophilic and to microroughen the surface by solvent and/or etching processes.

2. Electrodeposition

In this technique the metal to be coated is immersed in a conducting solution containing a salt of the coating metal and is made the cathode by applying an e.m.f. from an external source. The process of electrodeposition is essentially the cathodic site of the same electrochemical reaction which causes corrosion on the anodic site. The reaction is carried out under controlled conditions of electrolyte composition, potential and current density, and selected to favor the cathodic reduction of metal ions, so that metal is deposited rather than anodically oxidized to form metal cations or other oxidized species. Electrodeposits may be of pure metals, mixed metals, alloys, or metals mixed with nonmetallics.

Coating deposition initiates by nucleation at defects in the crystal lattice of the substrate metal, such as dislocations at the surface, with subsequent crystal growth of the deposited metal from the nucleated sites. By this mode of growth an adherent crystalline metal coating is built upon the substrate, bonded to it by atomic linkages.

3. Thermal Spraying

Sprayed metal coatings are obtained by making the coating metal molten and converting it into atomized globules. The molten globules are propelled to the substrate, are flattened on striking, and adhere to it. As successive globules strike and flatten on the surface they become partially welded together and a cohesive coating is built-up. The flattened solidified globules of coating adhere to the substrate surface purely by mechanical forces and there is no alloying action between the coating and the substrate. For this reason it is essential that the surface of the substrate should be clean and of a sufficient degree of roughness to provide adequate mechanical keying between coating and substrate. This is achieved by carefully controlled grit-blasting immediately prior to coating.

The thermal energy necessary to melt the spraying material can be produced in different ways. Flame, electric arc, and plasma as well as the Detonation Gun® are at present the most widely applied energy sources utilized in thermal spraying. Table 1 lists some of the coating properties as produced by various thermal spray techniques.

In flame spraying the coating metal is fed into a spray gun and melted by an oxyacetylene or oxypropane flame. The molten metal is atomized by the action of a stream of compressed air, together with the streaming effect of the heating flame itself, and propelled from the gun toward the substrate.

In electric arc guns the coating metal is supplied in the form of two wires carrying the current supply, the arc being struck at the point of contact. As with flame spraying, the molten metal is atomized and propelled by means of a carrier gas.

Plasma spraying resembles arc spraying in that a DC electric arc is employed for melting and atomizing the feed material, but in this case the arc is an ionized gas plasma struck between water-cooled metal electrodes.

The Detonation Gun® process uses measured quantities of oxygen, acetylene, and particles of coating material, detonated by a timed spark in a firing chamber. The molten particles, leaving the gun at high velocity, produce a high density layer on impact. The high kinetic energy of the particles is approximately 25 times that of the energy produced by flame spraying.

4. Chemical Vapor Deposition

It is possible to distinguish between three conceptually different types of treatment. The first type is that where an element such as carbon is allowed to diffuse into the surface to form an alloy with the substrate as in carburizing. Other examples of such diffusion profiles are chromizing, boronizing, and siliciding. The second type concerns the growth of a chemically distinct layer on the substrate, as in the case of titanium carbide. The third type concerns a combination of growth and reverse diffusion where, for example, chromium is deposited onto the surface and reacts with carbon which diffuses to it from the substrate to form a chemically distinct layer of chromium carbide on the surface. The mechanical response of the coated material can depend very much on the type of treatment given; in particular the adhesion of the coating depends on its mode of formation and on the nature of the interface.

The most commonly used treatment is the second type mentioned above. A simple definition of conventional chemical vapor deposition (CCVD) is the condensation of a compound or compounds from the gaseous phase onto a substrate where reaction occurs to produce a solid deposit. The process is shown schematically in Figure 1.

There are several principal types of thermochemical reactions in CCVD and these are loosely classified in the literature as follows:

1. Thermal decomposition: $W(CO)_6(g) \rightarrow W(s) + 6CO(g)$
2. Reduction: $WF_6(g) + 3H_2(g) \rightarrow W(s) + 6HF(g)$
3. Displacement or exchange: $SiCl_4(g) + CH_4(g) \rightarrow SiC(s) + 4HCl(g)$

Most chemical reactions in CCVD are endothermic, so that thermal energy must be supplied by heating the substrate and/or by the environment in the neighborhood of the substrate. Generally this is an advantage, since the reactions can be controlled by regulating

Table 1
COMPARISON OF SPRAYING TECHNIQUES

Process	Thickness	Typical coat	Porosity (%)	Advantages	Disadvantages
Flame	50 μmm to 3 mm	—	10—15	Fairly economic	Can get substrate heating high porosity; prior treatment of substrate
Arc	—	—	10—15	Greater bond strength than flame spraying; high deposition rates	Prior treatment of substrate high porosity
Plasma	—	Mo, Ni, Al_2O_3	1—10	Higher melting point materials can be used; high bond strength; heat source is inert, therefore, less oxidation	Costly
Detonation Gun®	—	Ni + Co, Al_2O_3, TiO_2	0—1	High hardness; very high bond strength	Costly

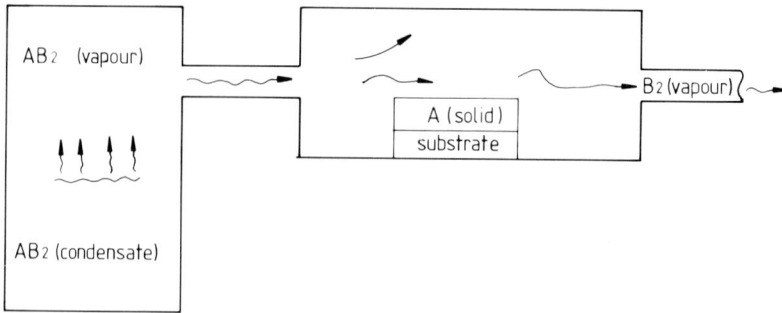

FIGURE 1.   Schematic representation of a CCVD process.

Table 2
SUMMARY OF CCVD
COATING SYSTEMS

Substrate	Coating	Substrate temp (°C)
Cu	Al_2O_3	450
Steel	TiB_2	>800
Steel	TiC	>800
Steel	Cr_3C_2	550
Steel	W_2C	800

the substrate temperature. However, this is a disadvantage in that only certain metals/alloys can be successfully treated by this method. In fact only systems which do not undergo phase transformations within the temperature range required for deposition can be employed. At present there is a very limited number of coatings which can be deposited below 600°C. Table 2 summarizes some CCVD coating systems.

5. Physical Vapor Deposition (PVD)

There are three basic PVD coating processes: evaporation, ion assisted, and sputtering. In the evaporation process, vapors are produced from a material located in a heated source. Typical energy sources utilized are radiation, electron beam bombardment, and electrical discharge. Deposition is normally conducted at pressures sufficiently low so that evaporated atoms undergo an essentially collisionless (line-of-sight) transport to the subtrate. Figure 2A is a schematic representation of a vacuum evaporation system. To overcome the problem of variation in coating thickness across the substrate, it is necessary to move the object being coated during the deposition process. A wide range of alloys (e.g., NiCr) and compounds (e.g., TiC) can be deposited. However, direct evaporation of compounds frequently yields deposits deficient in the nonmetallic species. Therefore, evaporation is often carried out in an atmosphere containing the deficient species in gaseous form, i.e., reactive evaporation.[1]

Should a plasma discharge be maintained in the atmosphere to promote the reaction between the gaseous and metal evaporated, then compounds can be deposited. This process of ion assisted evaporation is known as activated reactive evaporation (ARE)[2] and is shown schematically in Figure 2B. Another ion assisted process is known as ion plating. This terminology is applied to a coating process in which the substrate surface and/or the depositing film is subjected to a flux of high energy particles sufficient to result in improvements to interfacial strength and coating properties compared to a nonbombarded deposition. Figure 2C depicts a common design for ion plating equipment. One of the main advantages claimed

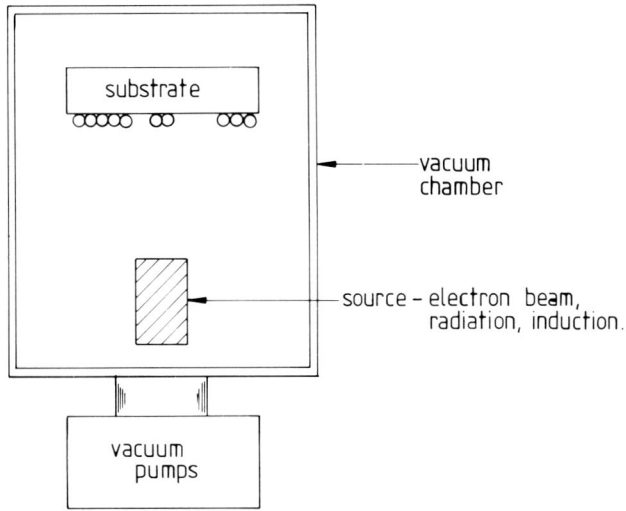

A

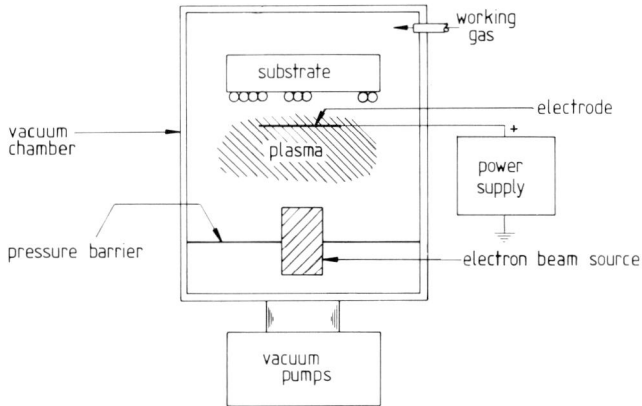

B

FIGURE 2.   Schematic representation of several PVD processes. (A) Evaporation; (B) activated reactive evaporation; (C) ion plating; (D) sputtering.

for ion plating is the very good adhesion obtained between coating and substrate, even when there is mutual insolubility. On the negative side, the act of high energy gas ions bombarding the surface, which leads to improved adhesion, results in a decreased deposition rate. A comparison of the deposition rates for the three main PVD processes can be seen in Table 3.

In the sputtering process, illustrated schematically in Figure 2D, the coating material is transformed to the vapor phase, primarily in the atomic form, by ion bombardment of a source electrode composed of the coating material. An important advantage of this process is the capacity to pass materials (metallic and ceramic) into the vapor phase while largely preserving chemical composition; even bone can be sputtered.[3] The deposition rates from sputtering tend to be lower than for other PVD processes, though developments over the past few years have meant that high rates are capable of being achieved (see Table 3). It is

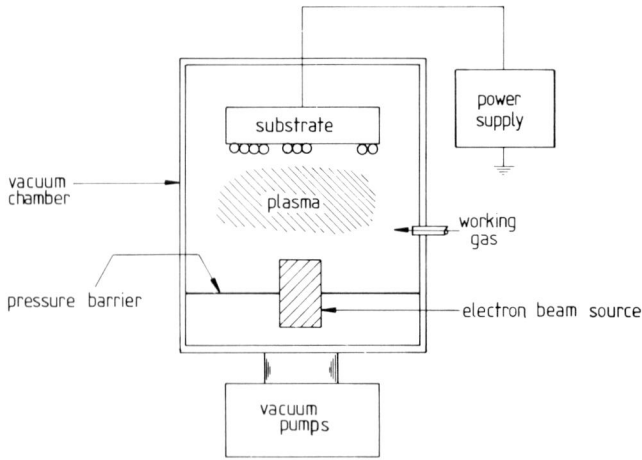

FIGURE 2C.

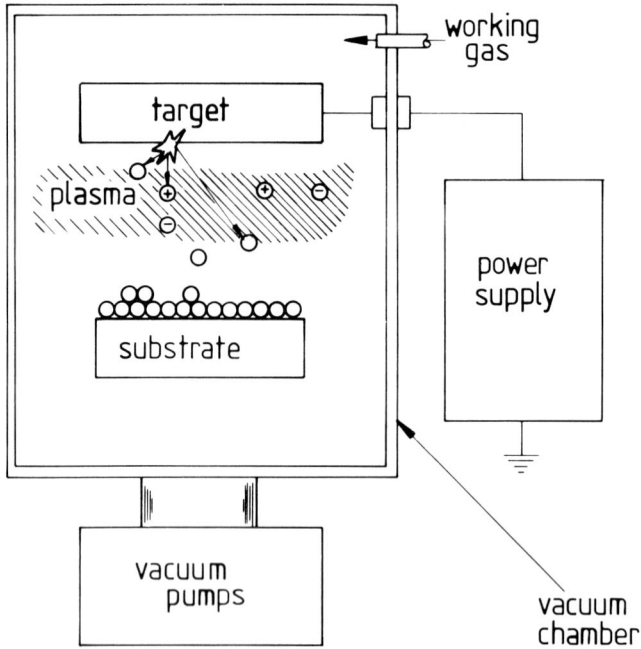

FIGURE 2D.

Table 3
A COMPARISON OF DEPOSITION
RATES FOR VARIOUS PVD
PROCESSES

Process	Deposition rate ($\overset{\circ}{A}$ min^{-1})
Evaporation	1×10^2—5×10^5
Ion plating	1×10^2—2.5×10^5
DC sputtering	25—25×10^2
Magnetron sputtering	0.5—3×10^4

Table 4
ADVANTAGES AND LIMITATIONS OF ION IMPLANTATION

Advantages	Disadvantages
Low temperature process	High capital cost
No toxicity	Requires *in vacuo* manipulation
Applied to finished components	(as do PVD processes)
Controllability	Line of sight process
	Shallow treatment

the development of sputtering sources with magnetic plasma confinement, called magnetrons, and of high performance triodes that has greatly enhanced the use of the sputtering process.

B. Surface Modification Methods

The processes described above have one common factor, that of depositing one material onto the surface of another. However, there are other techniques available which can induce property changes at the surface and for which a discrete surface layer interfacing with the substrate is not obtained. As mentioned earlier, ion implantation and laser glazing are two such surface modification processes.

1. Ion Implantation

The process of ion implantation provides a completely new method for improving the durability of material surfaces. Originally this technology was only used in the field of semiconductors; however, over the last decade, it has become useful in the metallurgical field. Ion implantation involves the bombardment of a target material with energetic ions (electrically charged particles). These ions penetrate the target to a depth dependent on the energy of the ion and the nature of the ion and the target. In a metallic substrate the neutralized ions come to rest in the lattice in interstitial and substitutional positions and remain there in a metastable solid solution. Thus, ion implantation may be used to overcome the equilibrium restraints on solid solubility associated with conventional alloying techniques.

The chosen atomic species are first ionized and then accelerated in an electric field to energies which lie in the range from a few tens to a few hundred kiloelectronvolts in a moderately hard vacuum. These ions arrive at the work piece (target) with velocities around that of a rifle bullet and can thus penetrate the surface. Such a process does not require a high temperature, the penetration being achieved by high kinetic energy.

In ion implantation there is no possibility of forming a distinct layer on the substrate. This is because the maximum concentration of implanted atoms is limited by sputtering (the kinetic ejection of target material under bombardment) to a level usually between 10 and 50 atomic percent. For the same reason there is almost no change in the dimensions of the workpiece. Like all processes ion implantation has advantages and disadvantages. Table 4 lists some of these points.

The apparent drawbacks of shallow treatment depth and the restrictions on the number of foreign atoms implanted may be overcome by research that is underway at the present. This new approach involves ion-beam induced intermixing of a thin coating and its substrate and the thermally activated process of radiation enhanced diffusion.

2. Laser Surface Treatments

Laser processing utilizes a laser to produce a source of energy which, when in contact with materials, can produce a large amount of heat in a very small volume of material. The amount of heat generated depends on the power density absorbed from the laser and the time of interaction with the surface. These parameters determine in which application the

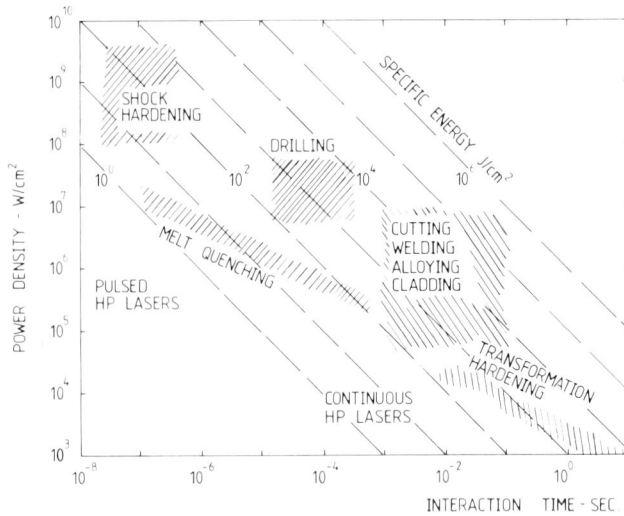

FIGURE 3.   Operational regimes for laser materials processing techniques.

laser will be used: welding, cutting, or surface transformation. It is the latter area which is of interest. Figure 3 illustrates the relationship between power density and the interaction time. Several mechanisms and treatments have been employed using lasers.

Laser transformation hardening relies on heat provided by surface absorption of the laser beam which is rapidly conducted into the cold core of the substrate. If the surface temperature of certain alloys, such as steel, is raised above a critical point and then cooled rapidly transformation will occur. This transformation is normally accompanied by an improvement in surface hardness. Laser transformation hardening systems are now widely used in the automobile industry.

If the beam power is sufficiently high and the material can absorb the incident radiation, local melting can occur: the power density is approximately three times that for transformation hardening. In laser alloying, constituents are added to this molten pool. The surface can thus be modified by melting in the alloying elements necessary for the end usage.

A recent development in which a surface liquid is produced has been described in the literature as laser glazing.[4] In this process a well-focused laser beam is scanned across the material at a rate which produces surface melting in localized regions while avoiding vaporization at the surface. Rapid surface melting occurs and an almost negligible amount of thermal energy is conducted into the base metal, thus producing temperature gradients between the solid and the liquid. Since there is always intimate solid/liquid contact, very rapid quenching of the melt to the cold, solid substrate results. As a result of this rapid chilling, a wide variety of unusual metallurgical microstructures can be produced; in fact amorphous layers can be produced at the surface of a bulk crystalline material.

Laser cladding involves the application of cladding (hard-facing) alloys to selected areas of a metal to produce improved surface characteristics. A precisely controlled laser beam melts only the cladding material, causing it to flow freely over the metal surface; freezing results in a homogeneous fine microstructural layer adhering to the substrate.

An alternative energy source to that of a laser is an electron beam. In fact the electron beam source is probably more versatile for the transformation hardening process than is a laser source (see Table 5). The hardening process works in exactly the same manner as when using a laser beam, i.e., localized heating of the surface, followed by solid-solid transformation. One novel and new application of electron beam technology is the consolidation of surface coatings.[5] The scanned electron beam provides an ideal localized heat

Table 5
COMPARISON OF ELECTRON BEAM AND
LASER BEAM HARDENING

Electron beam	Laser beam
Economic	Costly
(conversion of EB welders)	
Capital costs less	High capital
High power (60 kW)	Low power (20 kW)
No special surface coating	Certain metals require a special
	coating to minimize reflection
Vacuum process	Air process

source for the consolidation of coatings produced by processes such as plasma and flame spraying.

III. TECHNOLOGICAL APPLICATION OF SURFACE TREATMENTS

An ideal material for applications over a wide range of operating conditions should have at least the following properties:

1. Good mechanical properties.
2. Resistance to mechanical and thermal cycling.
3. Hardness and wear resistance.
4. Corrosion resistance.

In many applications, therefore, it is necessary to have a protective layer on the component so that the combined properties of the system can satisfy the operating conditions.

The choice of surface treatment for any given set of conditions depends on the operating environment, the substrate material, process availability, and economics. Thus, economics aside, the choice is dictated by the environment.

Because the requirements for a component facing a wear environment can be different from those in the corrosive or medical fields, these topics, i.e., wear, corrosion, fatigue, and medical, will be discussed separately. Although these examples describe uses outside the implant area, it may be seen that some of the findings do indicate specific improvements which might be made with similar materials used for implants

A. Tribological

Most wear problems fall into the general classes of (1) adhesive wear, resulting from the welding of surface asperities, and (2) abrasive wear, resulting from the cutting action of hard particles or asperities. Adhesive wear is often accompanied by the oxidation of wear debris which may in turn cause abrasive wear. Chemical inertness minimizes adhesive wear. Abrasive wear on at least one of the mating parts of a moving couple can be cured by making one of the surfaces harder than the abrasive. Table 6 shows some abrasive wear data for various surface treatment methods.

The use of solid lubricants is an effective method of eliminating adhesive wear. Films of direct current sputtered molybdenum disulfide (MoS_2) only 2000 Å thick have been found to have impressive lubricating characteristics and have an endurance that is more than 5 times greater than that of a resin bonded commercial film, 130,000 Å thick, when evaluated in a vacuum wear test.[6]

Ceramic materials, e.g., aluminum oxide (Al_2O_3) and titanium carbide (TiC), are particularly attractive for wear resistant coatings because of hardness and resistance to abrasive wear. Tool inserts offer an interesting example of combined adhesive and abrasive wear.

Table 6
ABRASIVE WEAR RESULTS FOR VARIOUS SURFACE TREATMENTS

Treatment	Test material	Coating	Thickness (μm)	Hardness (H_v)	Relative volumetric wear
CCVD	105 WCr6	TiC	12	3200	3.5
	C 100	W_2C	30	1900	4.8
	C 100	Borided	100	1600	12
Electroplating	Cu 40% Zn	Bright Ni	50	580	97
Anodizing	Al-Mg-Si	Natural oxide	20	450	63
Spray coating	Mo	Plasma	>100	750	33
	Al_2O_3	Plasma	>100	900	127

Effective coatings of titanium carbide can be sputter deposited and order of magnitude increases in the life of cemented carbide tool inserts, for cutting hardened 4340 steel,[7] have been found with this system. TiC coatings deposited by the ARE process have been found to improve the life of a high-speed tool steel by a factor of 300 to 800% when machining 4340 steel of hardness Rockwell C33.[8] In a sliding wear test, the wear rates of vacuum deposited chromium carbide and nitride coatings on stainless steel were lower than those of electroplated hard chromium coatings by a factor of 4 to 8 and by a factor of 300 than for the uncoated substrate.[9]

Ohmae et al.[10] have studied the durability of ion plated films under fretting wear since this wear process is regarded as one of the most complicated, involving adhesion, oxidation, abrasion, and fatigue.[11] A number of different materials were deposited on a 0.25% carbon steel. The investigation concluded that the ion plated gold coatings had a longer durability under fretting than those of vacuum deposited silver and gold (thermally evaporated) or ion plated silver coatings. Adhesion tests revealed that the adhesive strength of an ion plated gold film was higher than that of a sputtered film or a vacuum deposited film.

Introducing nitrogen into the glow discharge during ion plating can lead to the formation of nitride deposits.[12] Of the nitrides of the transition metals, titanium nitride is extremely chemically stable. Zega et al.[12] found that gold colored films of titanium nitride could be deposited onto various substrates, such as stainless steel and stellite, by reactive ion plating. The hardness of the coating tended to be low if the substrate temperature was only 400°C, but if it was increased to 800°C, the hardness was found to be as high as 2000 H_v.

If nitrogen alone is present in a glow discharge, no evaporation taking place, then nitriding of the substrate can occur. This process is known as ion nitriding. It is a similar process to conventional gas nitriding. Low alloy steels are treated by this process for automobile components. Taking these processes one stage further, i.e., implanting ion species directly into the surface, leads to a process where there is no build-up of material on the surface and no change in the component dimensions; this is ion implantation.

Ion implantation can be used to improve resistance to wear in a number of materials such as steel, titanium, and cemented carbides. The introduction of species such as nitrogen and carbon into steel brings about a hardening effect in the surface layers and has reduced the Archard wear parameter (volume of material eroded per unit sliding distance per unit load) by factors ranging between 20 and 30.[13] By way of example, Figure 4 shows the effect of nitrogen implantation on the volumetric wear rate of a stainless steel pin wearing against an implanted low alloy steel. Even at high loads the wear rate of the implanted component is reduced by an order of magnitude.[13]

Hirvonen[14] investigated the effect of ion implantation on the sliding wear of type 416 stainless steel (12% Cr — martensitic type) and AISI 52100 steel (1% C/1.5% Cr). A factor

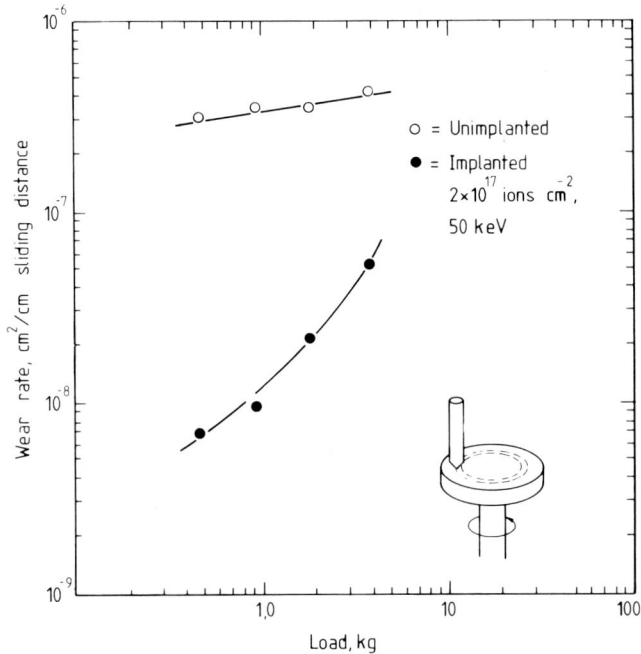

FIGURE 4. Volumetric wear parameter for a steel pin in contact with a disk of steel as a function of load.[13]

of two improvement in the wear of the AISI 52100 steel was found when nitrogen was implanted. However, improvements as high as a factor of 25 to 50 were found when nitrogen was implanted into the stainless steel.

Recently Bunshah and Suri[15] published data comparing the effect on wear of various surface treatments applied to 304 stainless steel, titanium, and aluminum. The treatments investigated were ion implantation and biased activated reactive evaporation (a process similar to ion plating). In the case of adhesive wear, ion implantation resulted in an improved wear behavior under lubricated conditions, but had no beneficial effect for dry wear conditions. Surface coatings, however, resulted in improved wear behavior for both the dry and lubricated conditions. For abrasive wear, it was observed that surface treatments were beneficial; see Table 7.

B. Corrosion

Metallic corrosion can be defined as the passage of a metal into the chemical combined state. The existing methods of corrosion prevention and control are necessarily based on attempts to interfere with the basic electrochemical mechanisms. Corrosion control techniques fall into five broad categories: materials selection, environmental control, electrochemical protection, the use of surface treatments, and design. It is the control of corrosion by surface treatment which will be discussed here.

Chemical vapor deposition (CVD) coatings have been used extensively as protective systems in various industries. Titanium fasteners, as used in the aerospace industry, are coated with aluminum; the use of CVD avoids hydrogen embrittlement that may result from conventional electroplating.[16] Ion plating has also been used to coat titanium components with aluminium.[17]

The long-term high temperature use of titanium alloys is limited by the high chemical reactivity of titanium. Fujishiro and Eylon[18] have investigated the ion plating technique in an effort to improve the high temperature properties of titanium alloys. They were able to

Table 7
COMPARISON OF SURFACE TREATMENTS ON FRICTION AND WEAR OF STAINLESS STEEL AND TITANIUM[15]

| Material | Surface treatment | Coefficient of friction | | Hardness (kg/mm²) | Abrasive wear | |
		Dynamic	Lubricated		Vol lost (1st hr)	Vol lost (2nd hr)
Stainless steel	None	0.42	0.15	296	0.53	0.30
(440 C)	N⁺ (implant)	0.17	0.15	632	—	—
	B⁺ (implant)	0.38	0.15	411	0.55	0.50
	TiN (coat)	1.15	0.225	2600	—	—
Titanium	None		0.7	229	1.95	1.35
	N⁺ (implant)	0.87	0.82	297	0.32	0.34
	B⁺ (implant)	0.7	0.85	330	—	—
	TiN (coat)	0.92	0.175	—	0.23	0.36

Note: 440 C stainless steel — martensitic type .

[a] Abrasive wear — using 600 grit SiC.

demonstrate that platinum deposits increased the creep resistance, and increased high cycle fatigue strength of a titanium alloy (Ti-6A1-2Sn-4Zr-2Mo); the high temperature oxidation resistance was also greatly improved.

Iron alloys, especially those containing more than 8 atomic percent chromium, can display extremely high resistance to pitting corrosion in chloride environments provided the alloy structure is amorphous.[19] One method of creating an amorphous structure is by high vacuum deposition techniques which offer a method of very fast quenching rates from the vapor to the solid phase. Nowak et al.[20,21] have studied the corrosion behavior of ion plated Fe-Cr films and found that the ferrous films containing at least 10% chromium can be crystalline, amorphous, or microcrystalline/amorphous depending upon alloy composition and deposition conditions. It was found that the ion plated films tended to exhibit increased rest potentials over the bulk material and a higher resistance to pitting in a neutral 1 *M* NaCl aqueous environment. Microcrystalline/amorphous films based on 304 stainless steel and produced by biased radio frequency sputtering have been found to exhibit very high pitting breakdown potentials and a greater passive region compared to the bulk material.[22]

It is possible to produce a material of markedly improved corrosion resistance by making appropriate alloying additions, e.g., additions of nickel and chromium to iron (i.e., stainless steels). Further, since corrosion is a surface phenomenon, it is not necessary to produce expensive bulk alloys to obtain an improved resistance to corrosion. In general improved resistance by alloy additions is only afforded by those additions that remain in solid solution in the base metal. In this context, ion implantation, to which the restrictions deriving from equilibrium phase diagrams do not apply, has the advantage that it offers the possibility of forming novel, metastable solid solution surface alloys, which are not accessible by conventional metallurgical techniques

Tantalum exhibits outstanding resistance to aqueous corrosion in a variety of environments and over a wide pH range. However, tantalum is insoluble in iron and cannot be considered as a conventional potential alloying addition. Work by Ashworth et al.,[23] using iron as a model system, demonstrated that implantation of tantalum produced a marked improvement in corrosion resistance. Tantalum implantation resulted in a significant decrease in both the critical current density for passivation and the passive current density when potentiokinetic polarization studies were performed in sodium acetate (acetic acid buffer solution of pH 7.3).

Table 8
FATIGUE LIFE OF UNCOATED AND
PLATINUM-COATED TITANIUM ALLOY
(Ti-6Al-2Sn-4Zr-2Mo) AT ROOM
TEMPERATURE

Surface condition	Maximum stress (MN/m²)	Number of cycles to failure
Uncoated	951	6.3×10^6
	703	2.6×10^6
	689	Run out
Coated	951	1.9×10^5
	827	1.3×10^7
	800	Run out

Other "passive film formers", such as chromium, can also be successfully implanted into less corrosion resistant materials. When chromium is implanted into iron, it has been found to behave electrochemically in the same manner as in the bulk alloy.[24] The test conditions used in this investigation were 20 keV Cr^+ implanted at a fluence of 5×10^{16} ion/cm and 2×10^{17} ions/cm; this produced Fe-Cr surface alloys of approximately 4 and 6%, Cr, respectively.

Needham et al.[25,26] have studied the effect of chromium and nickel implantation into iron and have concluded that the corrosion characteristics of the implanted alloy were substantially the same as those of bulk iron alloys, under the same experimental conditions. The implantation of chromium into a commercial Fe/18% Ni/9% Co maraging steel demonstrates the potential advantage of ion implantation for improving the corrosion behavior of a material without impairing its bulk properties.[26] In this particular case, applying a conventional protective chromium layer would require unacceptably long times if one used a processing temperature which did not produce significant grain coarsening, which in turn has detrimental effects on strength and toughness.

C. Fatigue

The term fatigue applies to the behavior of a metal which, when subjected to a cyclically variable stress of sufficient magnitude, produces a detectable change in mechanical properties. It is now well established that, in most cases, fatigue starts at the surface. Accordingly any alteration in surface properties must bring about a change in the fatigue properties. Fatigue resistance can be obtained in materials with a worked surface layer produced by processes such as shot peening or skin rolling the surface.

Fujishiro and Eylon[18] proposed that the increase in fatigue life of an ion plated titanium alloy was due to compressive hardening of the surface. The system investigated was a platinum coated Ti-6Al-2Sn-4Zr-2Mo alloy: fatigue testing was conducted using axial loading. Selected fatigue data are given in Table 8.

Stainless steels are known to be susceptible to corrosion fatigue particularly if in the sensitized state. Under these conditions it has been found that molybdenum coatings, deposited by the ion plating process, have lengthened the lifetime of fatigue samples.[27]

In 1970 Thompson[28] suggested that fatigue properties might by improved by ion bombardment because the movement of dislocations should be impeded by radiation damage. Compressive stresses are generated as in the process of shot peening, but because the effect is on an atomic scale, the troublesome surface irregularities which result from shot peening are eliminated.

Hartley[29] has reported that fatigue lifetimes for nitrogen implanted stainless steel, titanium

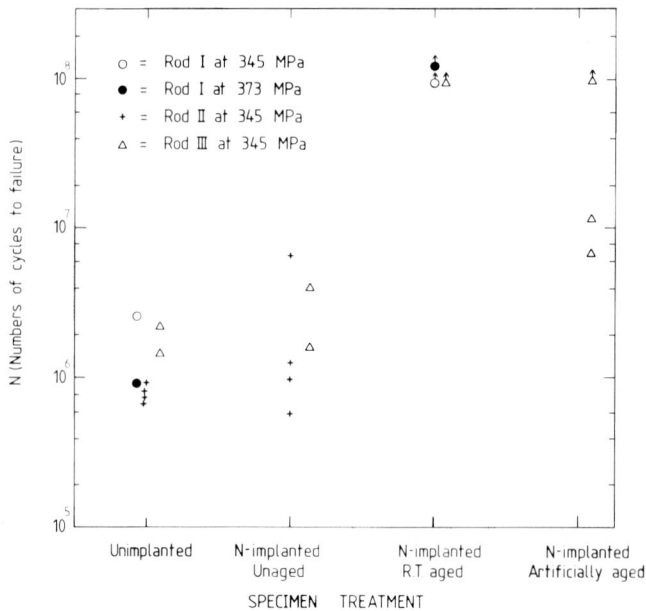

FIGURE 5.   Fatigue lifetime (cycles to failure) for AISI 1018 steel samples having different ion implantation treatments.[30]

and a maraging steel are 8 to 12 times longer than for the unimplanted materials (dose 2×10^{17} ions/cm). These results demonstrate that implantation of nitrogen can increase fatigue lifetimes. A more detailed study of this effect has been performed by Hu et al.,[30] who studied the effect of nitrogen implantation of 150 keV on the fatigue life of an AISI 1018 steel (1.2% C, 0.8% Mo). Following implantation the specimens were either tested promptly, aged at room temperature, or aged at a slightly elevated temperature. Early results showed no consistent improvement in fatigue life for samples freshly implanted. However, samples which had been aged at room temperature for several months prior to testing showed a measurable increase in life. This aging effective was confirmed by artificially aging at 100°C for 6 hr. The data obtained are summarized in Figure 5.

This work prompted more research into improvements in fatigue properties as a result of ion implantation. In another study, Hirvonen et al.[31] implanted both nitrogen and carbon, to a dose of 2×10^{17} ions/cm at 75 keV, into a Ti-6A1-4V alloy and found that the fatigue resistance was improved, particularly with implanted carbon. In fact an increase of almost 96 MNm^{-2} in the endurance limit was found.

Hochman and Marek[32] have studied the effect of implantation of molybdenum ions and the ion plating of molybdenum, platinum, tantalum, and titanium on the corrosion and fatigue properties of medical grade stainless steel. The investigation consisted of performing rotating bending and three point bend fatigue studies and conventional electrochemical polarization studies in Ringer's solution. It was observed that a coating of molybdenum had the most beneficial effect on fatigue life. Titanium and tantalum showed a less marked improvement, whereas platinum coated samples gave erratic results and tended to show a decrease in fatigue life. Similar effects were obtained when electrochemical passivation studies were performed. It was suggested that although a crack might develop or wear could remove some of the film, in the case of molybdenum serious corrosion would not occur because anodic protection would then safeguard the underlying metal.

IV. MEDICAL APPLICATIONS OF SURFACE TREATMENTS

The aim of using biomaterials is to restore the function of natural tissue and organs in the body. These aims can be met by the use of surface treatments such as the case with carbon coatings for cardiovascular prostheses. It has already been shown that surface treatments can improve wear, corrosion, and fatigue resistance. For use in the medical area these three properties are important, as is biocompatibility.

A. Carbon

During the last 10 years, carbon has been accepted as a material for use in the construction of implantable prosthetic devices. The various forms of carbon that have received attention are graphites, vitreous or glassy carbons, pyrolytic carbons, and vacuum vapor-deposited carbon coatings. Among these, the pyrolytic and glassy carbons have been studied in detail and have found clinical applications in cardiovascular, orthopedic, and dental areas as well as long-term percutaneous applications. This is a result not only of the biocompatability and inertness demonstrated by these materials, but also of the fact that the properties can be controlled by the processing parameters.

There are two main methods by which carbon is deposited, (1) pyrolysis of a hydrocarbon gas and (2) vacuum deposition. By using such techniques, three types of turbostratic carbons with distinctive characteristics can be prepared:

1. Pyrolytic carbons
2. Glassy (vitreous) carbons
3. Vapor-deposited carbons

Pyrolytic carbons are deposited in a fluidized bed from a hydrocarbon containing gaseous environment, such as methane, at temperatures in the range of 1000 to 2400°C. The carbon is deposited onto a suitable preformed substrate such as graphite. Although a wide range of structures is possible, only the dense, isotropic structures deposited at temperatures below 1500°C have proved useful in bioengineering. These are the low temperature isotropic or LTI carbons.

The glassy carbons are formed by controlled heating of a solid, preformed polymeric body to drive off volatile constituents, leaving a glassy carbon residue. The bodies are isotropic and monolithic, the thickness of the glassy carbons being limited to about 7 mm.

The main drawback with the pyrolytic carbons is the high deposition temperature. Vacuum processes allow deposition at much lower temperatures. Numerous PVD processes have been tried: ion plating, magnetron sputtering, and vacuum evaporation. Typical substrates for these systems are stainless steel and titanium.

By controlling the deposition conditions, in a fluidized bed, e.g., temperature, composition, time, it is possible to control the anisotropy, density, crystallite size, and the microstructure of the carbon deposited. Using the fluidized bed process, it is possible to introduce various other elements into the gas and codeposit these with carbon. The most useful addition is silicon.[33] The silicon is present as a dispersion of silicon carbide particles. The properties of various carbon materials are in Table 9.

An additional property of substantial importance for some prosthetic devices is low friction for carbon-carbon, carbon-plastic, and carbon-metal combinations. Listed in Table 10 are the volume wear rates for various combinations of LTI carbon, coupled with UHMWPE and metals. Briefly, the wear apparatus consisted of a polished disk with a controlled curvature which rotated at 60 r/min against a polished, flat test specimen; the fluid medium was distilled water (at room temperature). Under comparable conditions the difference in volume wear rate among the materials studied was small.

Table 9
COMPARISON OF PROPERTIES OF VARIOUS CARBONS[65]

Property	Glassy carbon	Vapor deposited carbon	LTI carbon	LTI carbon with silicon[a]
Density, g/cm³	1.4—1.6	1.5—2.2	1.7—2.2	2.04—2.13
Flexural strength, MN/mm²	69—207	345—700	276—552	552—620
Young's modulus, GN/m²	24—31	14—21	17—27	27—31
Hardness, DPH[b]	150—200	150—250	150—250	230—370

ᵃ   5.0 to 12.0 wt % Si.
ᵇ   Diamond pyramid hardness.

Table 10
VOLUME WEAR RATES OF VARIOUS
MATERIAL COMBINATIONS[66]

Material combination (disk/flat)	Volume wear rate (10^{-6} mm³/km)
LTI carbon/Ti	1.16 ± 0.07
LTI Carbon/Ti-11.5 Mo-6Zr-4.5Sn	1.46 ± 0.08
LTI carbon/UHMWPE	1.47 ± 0.23
Stellite 21/UHMWPE	1.56 ± 0.30
LTI carbon/LTI carbon	1.23 ± 0.06

One of the most demanding of applications for a biomaterial is in artificial heart valves. Not only must the material interface with blood, but it must be able to endure 10^9 or more stress cycles during the anticipated life of a patient without degrading, wearing out, or failing in fatigue. Some valves currently utilize pyrolytic carbon coatings for all blood contacting surfaces. In 1969 De Bakey introduced the first clinical aortic valve using a hollow pyrolytic carbon-coated ball; these valves also utilized carbon-coated metal struts. Because of its demonstrated biocompatibility, durability, inertness, and immunity to dynamic fatigue, other clinical valves have been developed that use LTI carbon components. The occluders in the Beall-Surgitool (mitral) valve, the Bjork-Shiley (mitral and aortic) valves, and the Cooley-Cutter (mitral and aortic) valves are all fabricated from LTI carbon.[33]

Since the introduction of carbon for use in cardiovascular prostheses, considerable effort has been expended to produce all-carbon prosthetic heart valves, e.g., in the Omnicarbon®* valve. The sewing ring of this prostheses is also vacuum coated with carbon.

Pyrolytic and vitreous carbons have also been studied since 1971 as a possible artificial tooth root,[34] i.e., an implantable support for a single tooth, a bridge, or a complete denture. From clinical and animal studies, the shape and size of a dental implant in relation to its osseous supporting structure are important factors and, for a given site, removal of more than a critical amount of tissue can reduce the probability of achieving a successful implant. There are many situations that require that an implant be smaller than can be manufactured using a pyrolytic carbon. In these cases, vapor-deposited carbon coatings can be applied to substrates such as Vitallium®** (Co-Cr-Mo) alloy. Such an implant will behave mechanically as a metal, but chemically as a carbon. Leake et al. have evaluated vapor-deposited carbon as a coating for cast metallic implants fabricated for subperiosteal dental implants[35] and for

*   Medical Incorporated, Minneapolis, Minn.
** Howmedica Incorporated, New York.

endosseous tooth root replicas.[36,37]Cranin et al. have studied carbon-coated Vitallium® dental implants in animals and humans.[38-40] Results have been encouraging.

Vacuum deposited carbon has also been used to advantage for artificial arteries.[41] The most popular synthetic material used for large artery replacement is loosely knitted Dacron®.* However, no satisfactory synthetic material has proved to be suitable for the replacement of arteries smaller than 6-mm diameter. By combining carbon and Dacron®, it was found that 4-mm-diameter carbon coated Dacron® grafts could be fabricated. The study showed that the new graft was, at the least, equal in performance to the autologous vein grafts used for arterial replacement after implantation of 21 days in the canine carotid artery.

Although the unique combination of biochemical and biomechanical properties of carbon is being used to advantage clinically in many areas such as prosthetic heart valves and tooth roots, it has also been investigated for use in the area of orthopedics.

Several artificial joints have been manufactured from LTI carbon, e.g., a tibial plateau for an artificial knee, a femoral head and acetabular cup, a finger joint, and a bone plate and screws.[42,43] McKibbin[44] has also investigated the use of a carbon coating for fracture fixation devices. It was found that during the use of this plate in an open wound situation it behaved biochemically as pure carbon. The carbon-coated stainless steel plate was epithetialized with the same ease as a pure carbon plate, thus obtaining the biological advantages of carbon without sacrificing any mechanical properties. No breakdown in the coating was observed after 6 months implantation in sheep.

B. Surface Coatings

Vapor deposition systems have also been used for neurological electrodes. Coaxial depth electrodes for implantation in the CNS are of interest for basic research and diagnostic studies of neurological diseases. Reduction of the total diameter of these electrodes is desirable to minimize tissue trauma during implantation. Fine electrodes of diameter 0.2 mm (0.008 in.) were fabricated by using both vacuum evaporation (PVD) and CVD.[45,46] Coaxial electrodes consisting of titanium conductors with titanium dioxide insulator in between were fabricated by PVD techniques. Coaxial tungsten conductors with a mixed alumina-silica insulating layer in between were fabricated by CVD techniques.

In an attempt to combine the advantages of metals (toughness) and ceramics (biological compatibility and corrosion resistance), coating of metallic devices by flame and/or plasma sprayed ceramics has been performed.[47-50] The resultant metal-ceramic configuration consists of a thin, slightly porous coating bonded to a metal core. By controlling the flame-spraying application parameters, the porosity of the coating can be regulated, whereby a porous ceramic coating capable of accepting permanent bone or tissue ingrowth can be fabricated. Clinical studies have been undertaken to evaluate alumina bonded to metal.[48,49] The results of this work have not been encouraging for applications in orthopedics; the strength of the flame sprayed ceramic-to-metal bond is low and not strong enough for load bearing applications (see Table 11).

Bortz and Onesto have developed prosthetic devices using methods and materials developed for space.[51] Free standing metal reinforced ceramic tubes have been fabricated using flame and plasma spraying techniques. It was found that these materials showed good body compatibility as trachea and vena cava implants. In developing an artificial trachea, a layered structure consisting of 254 μm of flame sprayed zirconia, 254 μm of flame sprayed stainless steel (type 316), and 254 μm more of zirconia was fabricated on a polished copper-coated mandrel. No prostheses were rejected after 30 days of implantation into rabbits. Also several sprayed metal-ceramic composite tubes were studied for compatability with blood in order to evaluate them for use as vein prostheses. For this the tubes were evaluated by implanting

---

* Du Pont de Nemours, Wilmington, Delaware.

Table 11
MAXIMUM BOND STRENGTH FOR FLAME-SPRAYED ALUMINA ON STAINLESS STEEL[48]

Surface treatment	Bond strength, max MN/m^2	Failure location
Anodic polarization of metal surface	73	Ceramic-metal interface
Machine roughening of metal surface	3.5	Ceramic-metal interface
Alumina grit blast of metal surface	1.3	Ceramic-metal interface
3 Weeks of immersion in aerated Ringer's solution (metal as Treatment 1)	2.3	—

them in vena cavae of dogs. No clotting was observed during pathology studies conducted at sacrifice, the maximum length of implantation being 6 weeks.

It has not always been found necessary to use technology as advanced as that developed for space applications. Colmano et al.[52] have used stainless steel electrodes electroplated with silver to inhibit the in vitro growth of *Staphylococcus aureus* in rabbits. This investigation was based on the knowledge that a silver electrode (anode) activated by 0.4 to 400 μA of positive direct current per cubic millimeter exhibits a bacteriostatic effect on *S. aureus*.[53,54] Evidence from these experiments indicated that the observable inhibition of a stainless steel electrode electroplated with silver was in the region of 85 to 90% bacterial control.

Electroplating has also been used in other medical areas. Hard chrome plating of a Co-Ni-Cr-Mo wrought alloy (Protasul-10®*) has been used for applications in total hip prostheses. In an investigation concerned with the wear characteristics of Charnley-Müller hip joint endoprostheses, it was found that a hard chrome plated Protasul-10® alloy behaved no differently than a cast Co-Cr-Mo alloy (Protasul-2)® when running against UHMWPE in vivo.[55-57] However, the chromium plated Protasul-10® head has not been used since 1975, the company reverting to a Protasul-2® head which is now attached onto a Protasul-10® stem.

Surface treatments do not always result, however, in beneficial properties. Uhthoff and Bardos found that a Ti-6Al-4V alloy with a nitrided surface was not acceptable for use in the treatment of fractures because of an unacceptable degree of tissue reaction.[58] This reaction was found to be due to flaking of the coating. A detailed examination revealed that the coating spalled when the substrate surface was put in tension.

In contrast to this observation several investigators have found that nitriding the surface of Ti-6Al-4V alloy does not produce a significant increase in the wear rate when in contact with UHMWPE.[59,60] In fact McKellop et al.[59] found that if abrasive particles of bone cement were added to the testing environment, the wear damage was minimal for the nitrided surface, whereas surface scratching began immediately with the untreated titanium alloy. Further work on creating a surface layer on titanium or its alloys to prevent abrasion, corrosion, and fretting wear has resulted in the granting of a patent to the Institute Straumann.[61] It has been claimed that a surface coated titanium implant possesses a greater resistance to abrasion than an implant untreated. The surface coating is comprised of one of the following group which includes an oxide, nitride, carbide, and carbonitride. The coatings can be applied by either (1) anodic treatments or (2) CVD.

---

* Sulzer Bros., Winterthur, Switzerland.

C. Surface Modifications

At the present time little has been published on the actual use of surface modification treatments in medical applications; though the potential does exist, e.g., the use of ion implantation with medical grade alloys.[32] What has been published has been concerned with modifying the surface morphology of biological implants.[62-64] In these studies an electron bombardment ion thruster (developed for space technology) was used as an ion-beam sputtering source to texture the surface of biomaterials. The materials investigated included 316 type stainless steel, Co-Cr-Mo alloy, Co-Ni-Cr-Mo alloy, silicone rubber, and a polyolefin. The objective of these investigations can be split into two areas, (1) soft tissue attachment and (2) hard tissue attachment. The first group, soft tissue implants, includes implants intended for vascular prostheses, artificial heart pump diaphragms, pacemakers, and percutaneous connectors. The second group of implants, the hard tissue prostheses, includes dental implants and orthopedic appliances. By texturing the surface with arrays of pores of differing geometries, an optimum implant surface texture for bone, soft tissue, or thrombus attachment can be fabricated. This work is still in progress; however, preliminary results indicated no serious change in tissue response.

V. CONCLUSION

Many different techniques are available for surface treatment of bulk materials. Broadly speaking these techniques either employ coating or surface modification techniques in order to improve properties such as fatigue, corrosion, or wear resistance. Impressive improvements in properties have been demonstrated in several cases and surface treatment techniques are being employed in some implant applications. It is believed that increased use of such techniques will be made as the potential for property improvements becomes more widely recognized.

GENERAL REFERENCES

a. **Carter, V. E.,** *Metallic Coatings for Corrosion Control,* Newnes-Butterworth, London, 1977.
b. **Maissel, L. I. and Clang, R.,** *Handbook of Thin Film Technology,* McGraw-Hill, New York, 1970.
c. Union Carbide, Coating Services Division, Indianapolis, Indiana.
d. **Dearnaley, G., Freeman, J. H., Nelson, R. S., and Stephen, J.,** *Ion Implantation,* North-Holland, Amsterdam, 1973.

REFERENCES

1. **Ritter, E.,** *J. Vac. Sci. Technol.,* 3, 225, 1966.
2. **Bunshah, R. F. and Raghuram, A. C.,** *J. Vac. Sci. Technol.,* 9, 1385, 1972.
3. U.S. Patent 2,918,100, 1975.
4. **Breinan, E. M., Banas, C. M., and Greenfield, M. C.,** Report No. AFML-TR-75-149, Ohio 1975.
5. **Tucker, T. R., Ayers, J. D., and Schaefer, R. J.,** in *Laser and Electron Beam Processing of Materials,* Academic Press, New York, 1980.
6. **Buckley, D. H. and Spalvins, T.,** Sputtering and Ion Plating, Report No. SP-5111, National Aeronautics and Space Administration, Houston, 1972.
7. **Carson, W. W.,** *J. Vac. Sci. Technol.,* 12, 845, 1975.
8. **Bunshah, R. F. and Shabaik, A. H.,** *Res. Dev.,* 26, 46, 1975.
9. **Nakamura, K., Inagawa, K., Tsuruoka, K., and Komiya, S.,** *Thin Solid Films,* 40, 155, 1977.
10. **Ohmae, N., Tsukizoe, T., and Nakai, T.,** *Wear,* 30, 299, 1974.
11. **Waterhouse, R. B.,** *Fretting Corrosion,* Pergamon Press, Oxford, 1972.
12. **Zega, B., Kornmann, M., and Amiquet, J.,** *Thin Solid Films,* 45, 577, 1977.

13. **Dearnaley, G. and Hartley, N. E. W.**, in Proc. of Ion Plating and Allied Techniques (IPAT), CEP Consultants, Edinburgh, 1977.
14. **Hirvonen, J. K.**, *J. Vac. Sci. Technol.,* 15, 1662, 1978.
15. **Bunshah, R. F. and Suri, A. K.**, in Proc. of Ion Plating and Allied Techniques, CEP Consultants, London, 1977.
16. **Blocker, J. M. and Withers, J. C., Eds.**, *Proc. 2nd Int. Conf. on Chemical Vapour Deposition,* Electrochem. Soc., New York, 1970.
17. **Teer, D. G. and Salem, F. B.**, *Thin Solid Films,* 45, 333, 1977.
18. **Fujishiro, S. and Eylon, D.**, *Thin Solid Films,* 56, 309, 1978.
19. **Naka, M., Hashimoto, K., and Masumoto, T.**, *Corrosion,* 32, 146, 1976.
20. **Nowak, W. B. and Collins, M. R.**, in Rapidly Quenched Metals, 3rd Int. Conf., Sussex, 1978.
21. **Nowak, W. B. and Okorie, B. A.**, in Proc. of Ion Plating and Allied Techniques, CEP Consultants. London, 1979.
22. **Nowak, W. B.**, *Mater. Sci. Eng.,* 23, 301, 1976.
23. **Ashworth, V., Baxter, D., Grant, W. A., and Procter, R. P. M.**, *Corros. Sci.,* 17, 947, 1977.
24. **Grant, W. A.**, 7th Int. Summer School and Symposium on the Physics of Ionized Gases, Yugoslavia, 1976.
25. **Sartwell, B. D., Campbell, A. B., and Needham, P. B.**, in *Proc 5th Int. Conf. on Ion Implantation in Semiconductors and other Materials,* Plenum Press, New York, 1977.
26. **Covino, B. S., Jr., Needham, P. B., and Conner, G. R.**, *J. Electrochem. Soc.,* 125, 370, 1978.
27. **Gregory, B. and Ozcan, R.**, *Eng. Med.,* 9. 3, 1980.
28. **Thompson, H. W.**, in Proc. Eur. Conf. on Ion Implantation, Dearnaley, G., Ed., Reading, U.K., 1970.
29. **Hartley, N. E. W.**, in *Applications of Ion Beams to Materials,* Carter, G., Collignon, J. S., and Grant, W. A., Eds., Institute of Physics, London, 1976.
30. **Hu, W. W., Clayton, C. R., Herman, H., and Hirvonen, J. K.**, *Scr. Metall.,* 12, 697, 1978.
31. **Hirvonen, J. K., Carosella, C. A., Kant, R. A., Singer, I., Vardiman, R., and Rath, B. S.**, *Thin Solid Films,* 63, 5, 1979.
32. **Hochman, R. F. and Marek, M.**, Paper presented at 1st World Biomaterials Congress, Vienna, 1980.
33. **Bokros, J. C., La Grange, L. D., and Schoen, F. J.**, *Chemistry and Physics of Carbon,* Vol. 9., Walker, P. L., Ed., Marcel Dekker, New York, 1972.
34. **Grenoble, D. E.**, *Ariz. State Dent. J.,* 19, 12, 1973.
35. **Leake, D., Michieli, S., Freeman, S., Bokros, J. C., and Haubold, A.**, Abstract B-4(2) in Extended Abstracts, 13th Biennial Conference on Carbon, Irvine, Calif., July, 1977.
36. **Michieli, S., Leake, D., Freeman, S., Bokros, J. C., and Haubold, A.**, Abstract B-4(3), in Extended Abstracts, 13th Biennial Conference on Carbon, Irvine, Calif., July, 1977.
37. **Michieli, S., Leake, D., and Pizzoferrato, A.**, Abstracts 1st World Biomaterials Congr., Vienna, April 1980, 231.
38. **Cranin, A. N., Rabkin, M. F., Silverbrand, H. S., and Sher, J. H.**, Trans. 4th Annu. Meet. Soc. of Biomaterials, San Antonio, Texas, April 1978.
39. **Kent, J., Cranin, A. N., Meffert, R., Armitage, J., Bubbush, C., James, R., Shim, H., and Bokros, J. C.**, Abstracts of 1st World Biomaterials Congr., Vienna, April 1980, 232.
40. **Satler, N., Cranin, A. N., Silverbrand, H. S., and Sher, J. H.**, Abstracts of 1st World Biomaterials Congr., Vienna, April 1980, 233.
41. **Sharp, W. V., Scott, D. L., and Teague, P. C.**, Trans. 4th Annu. Meet. Soc. of Biomaterials, San Antonio, Texas, April 1978.
42. **Haubold, A.**, *Ann. N.Y. Acad. Sci.,* 283, 383, 1977.
43. **Bokros, J. C., Akins, R. J., Shim, H. S., Haubold, A., and Agarwal, N. K.**, *Am. Chem. Soc. Symp,* No. 21., 1975, p. 237
44. **McKibbin, B.**, *Current Concepts of Internal Fixation of Fractures,* Uhthoff, H. K., Ed., Springer-Verlag, Berlin, 1980.
45. **Darolia, R., Bunshah, R. F., Martin, P. L., Dymond, A. M., and Crandall, P. H.**, 28th Annu. Conf. Eng. in Med. Biol., September 1975.
46. **Martin, P. L., Bunshah, R. F., and Dymond, A. M.**, *J. Vac. Sci. Technol.,* 12, 754, 1975.
47. **Hulbert, S. F., Young, F. A., Mathews, R. S., Klawitter, J. J., Talbert, C. O., and Stelling, F. H.**, *J. Biomed. Mater. Res.,* 4, 433, 1970.
48. **Baldwin, C. M. and MacKenzie, J. D.**, *J. Biomed. Mater. Res.,* 10, 445, 1976.
49. **Drummond, J. L., Simon, M. R., and Brown, S. D.**, Corrosion and Degradation of Implant Materials, ASTM STP 684, American Society for Testing and Materials, Philadelphia, 1979.
50. **Brown, S. D., Reed, D. P., Drummond, J. L., Ferber, M. K., and Simon, M. R.**, Trans. 4th Annu. Meet. Soc. of Biomaterials, San Antonio, Texas, April 1978.
51. **Bortz, S. A. and Onesto, E. J.**, *Composites,* 151, 1974.
52. **Colmano, G., Edwards, S. S., and Barranco, S. D.**, *Va. Med. Mon.,* 106, 928, 1979.

53. **Barranco, S. D., Spadaro, J. A., and Berger, T. J.,** *Clin. Orthop. Relat. Res.,* 100, 250, 1976.
54. **Spadaro, J. A., Berger, T. J., and Barranco, S. D.,** *Antimicrob. Agents Chemother.,* 6, 637, 1974.
55. **Kriete, U., Willert, H. G., and Semlitsch, M.,** Trans. 4th Annu. Meet. Soc. of Biomaterials, San Antonio, Texas, April 1974.
56. **Kriete, U., Willert, H. G., Semlitsch, M., Weber, H., Lehmann, M., Niederer, P. G., and Dörre, E.,** Plastics in Medicine and Surgery, 3rd Int. Conf., Twente, Holland, 1979.
57. **Willert, H. G., Buchhorn, U., and Zichner, L.,** *Arch. Orthop. Traumat. Surg.,* 97, 197, 1980.
58. **Uhthoff, H. K. and Bardos, D.,** Trans. of the 24th Annu. Meet. Orthopedic Res. Soc., Dallas, February 1978.
59. **McKellop, H. A., Clarke, I. C., Markolf, K. L., and Amstutz, J. C.,** Trans. of 5th Annu. Meet. Soc. of Biomaterials, Clemson, S.C., April 1979.
60. **Miller, D. A., Ainsworth, R. D., Dumbleton, J. H., Page, D., Miller, E. H., and Shen, C.,** *Wear,* 28, 207, 1974.
61. U.S. Patent 3,643,658, 1972.
62. **Weigand, A. J. and Banks, B. A.,** *J. Vac. Sci. Technol.,* 14, 326, 1977.
63. **Weigand, A. J.,** in Trans. of the 4th Annu. Meet. Soc. of Biomaterials, San Antonio, Texas, April 1978.
64. **Gibbons, D. F.,** personnal communication.
65. **Bokros, J. C.,** *Carbon,* 15, 353, 1977.
66. **Shim, H. S.,** *J. Bioeng.,* 1, 223, 1977.

Chapter 6

CALCIUM PHOSPHATE CERAMICS

W. Van Raemdonck, P. Ducheyne, and P. De Meester

TABLE OF CONTENTS

I. INTRODUCTION

The name "apatites" is given to a group of solids characterized by the chemical formula $M_{10}(XO_4)_6Z_2$. The apatites form a vast range of solid solutions; considerable variations among apatites can arise by ionic substitutions. Among the apatites, hydroxyapatite ($Ca_{10}(PO_4)_6(OH)_2$) is a scientifically important substance with considerable technological applications. This compound is found in nature and can also be synthesized. It is used extensively in the chromatographic separation of proteins. Another major interest for these apatites arises from the fact that hydroxyapatite crystals are the inorganic constituent of osseous and dental tissues and that synthetic hydroxyapatites are interesting biomaterials.

In this chapter the structure and composition of synthetic hydroxyapatite will be described first. The deviations from the ideal structure as a result of ionic substitutions will be discussed. In order to enhance the understanding of the behavior of synthetic hydroxyapatite, the structure and properties of other synthetic apatites, natural hydroxyapatite, and other calcium phosphates will be mentioned. Eventually all these data form the basis for the description of the in vivo behavior of man-made hydroxyapatites.

II. CHEMICAL COMPOSITION AND CRYSTAL STRUCTURE OF SYNTHETIC HYDROXYAPATITE

A. Chemical Composition

To obtain a clear understanding of the chemical composition of synthetic hydroxyapatite, the description of the composition of the general apatite group is useful. As was pointed out in the introduction, apatites form a family of solids with the general chemical formula $M_{10}^{2+}(XO_4^{3-})_6Z_2^-$. These apatites are frequently nonstochiometric; 1 mol of apatite may contain less than 10 mol of metallic ions at the M^{2+} positions and less than 2 mol of anions at the Z^- positions. However, the number of moles at the XO_4^{3-} position always remains 6. The M^{2+} ions are double charged cations like Ca^{2+}, Sr^{2+}, Ba^{2+}, Pb^{2+}, and Cd^{2+}. The XO_4^{3-} ions are anions such as AsO_4^{3-}, VO_4^{3-}, CrO_4^{3-} and MnO_4^{3-}, while the monovalent Z^- ions are F^-, OH^-, Br^-, and C_2^- ions.[1]

The present structure is not always realized. More complex ionic structures can occur. Instead of the monovalent anions, bivalent anions like O^{2-}, O_2^{2-}, and CO_3^{2-} can occupy the Z-positions.[2-6] When one bivalent anion replaces two monovalent anions, the electrical neutrality is preserved and at the same time one anionic position becomes vacant. Examples are the oxyapatite $Ca_{10}(PO_4)O_6\square$[7-10] and $Ca_{10}(PO_4)_6CO\square_5$ the so-called A-type carbonated apatite. The sign $\square$ stands for a vacant position.

The M^{2+} positions may also partially consist of vacancies. The neutrality of the crystal structure is preserved by two modifications that can occur simultaneously. The first modification is the appearance of vacancies at the Z-positions like in the nonstoichiometric hydroxyapatite $Ca_{9.8}\square_{0.2}(PO_4)_6(OH)_{1.4}\square_{0.6}$.

In the second modification, the same vacancies originate together with the substitution of some of the PO_4^{3-} ions by bivalent groups like in the carbonated apatite[2,5] $Ca_8\square_2(PO_4)_4(CO_3)_2\square_2$. This carbonated apatite of the so-called B type presents pronounced differences with the A type carbonated apatite, and the localization of the substituting CO_3^{2-} ions, respectively, at the PO_4^{3-} sites and the OH^- sites, can be clearly distinguished using IR spectroscopy and X-ray diffraction and by wet chemical analyses.[2,3]

The trivalent anionic groups may even be substituted by quadrivalent groups such as SiO_4^{4-} and GeO_4^{4-}. The electrical neutrality is then preserved by substitution or vacancies at the M^{2+} and Z^- positions. However, no examples are known of apatites with vacancies at the XO_4^{3-} sites.[1]

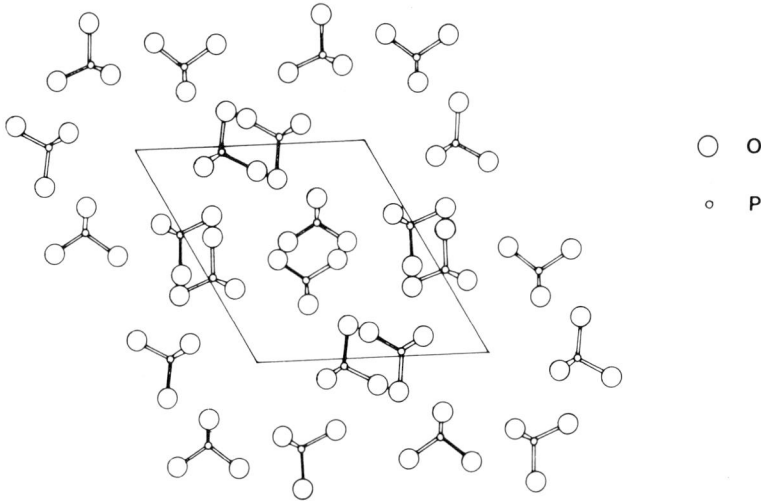

FIGURE 1.   Projection of the PO_4 groups of hydroxyapatite on the basal (001) plane.[11]

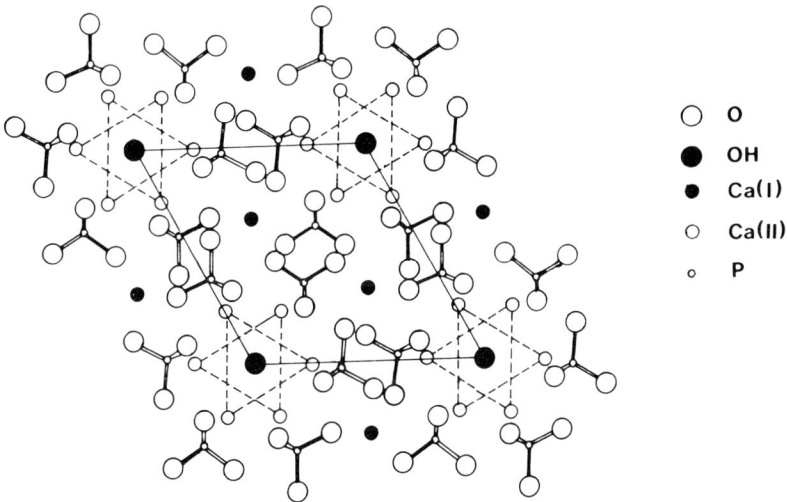

FIGURE 2.   Projection of the constituting ions of hydroxyapatite on the basal (001) plane[11]

B. Crystal Structure of Hydroxyapatites

The crystal structure of apatites is well known in the case of simple compositions which are obtained naturally or synthetically as monocrystals. We focus on the structure of hydroxyapatite. This is shown in Figures 1 and 2. It is a hexagonal structure with a $P6_3/m$ space group and with cell dimensions a = b = 9,42 Å and c = 6,88 Å.[11] ($P6_3/m$ stands for a space group with a sixfold symmetry axis with a threefold helix and a mirrorplane.) However, it is known that hydroxyapatite can occur in a closely similar pseudohexagonal monoclinic form with space group $P2_1/b$, if it is pure and free of vacancies.[11,12] $P2_1/b$ is a space group with a twofold rotation axis and an axial translation of width b/2 along (010).

Hydroxyapatite consists of a skeleton of PO_4^{3-} tetrahedra: two oxygen atoms are in a horizontal plane; the other two are on an axis parallel to the c-axis. In one crystal unit cell the PO_4^{3-} tetrahedra are divided in two layers, respectively, at a crystal height of 1/4 and 3/4. Figure 1 shows a projection of PH_4^{3-} groups on the basal (001) plane. The PO_4^{3-}

tetrahedra are distributed in that way so that two types of channels exist, perpendicular on the basal plane.

1.   The first channel, with a diameter of 2 Å, coincides with the ternary axis and is occupied by calcium ions, the co-called Ca(I) ions. In each unit cell, two channels are found, each of them occupied by two Ca ions; these ions lie upon a height of 0 and 1/2 of the crystal.
2.   The second channel with a diameter of 3 to 3.5 Å has a helicoidal sixfold symmetry axis. The walls of this channel consist of oxygen atoms and other calcium ions, the so-called Ca(II) ions. They are located at heights 1/4 and 3/4. They consist of two equilateral triangles rotated 60° in the plane perpendicular to the c-axis. The location of the Ca(I) and Ca(II) atoms is shown in Figure 2.

The second type channel plays an important role in the physicochemical behavior of the apatites.[1]

Various ions are known to be accommodated, apparently always on or very near the symmetry axis, in these channels, including F^-, Cl^-, OH^-, O^{2-}, CO_3^{2-}, and various combinations thereof. It is within these channels then that the distinction between the hexagonal and monoclinic form arises. The principal difference between the two hydroxyapatites is the ordering of the OH^- ions.[11,12] This is shown in Figures 3A and B. In the hexagonal form the OH^- ions are in twofold disorder about the mirrorplane, and each OH^- position is statistically only 50% occupied. Thus, in any particular local region, the OH^- ions have to be similarly oriented, otherwise the H ions would be required to come impossibly close together. Monoclinic hydroxyapatite has an ordered arrangement of hydroxyl ion columns and the b-axis is doubled.[13,14]

Based on the above description of the structure, localization of the substituting ions is now possible. The cationic substitutions or vacancies occupy the Ca positions. The monovalent anions OH^-, Cl^-, Br^-, and some bivalent anions, 0^{2-}, O_2^{2-}, S^{2-}, CO_3^{2-}, occur on the sixfold symmetry axis and depending upon their atomic diameter take place in the center of the Ca(II) triangles (like F^- and OH^- with small dimensions) or in the center of the oxygen triangles (larger substituting ions like Cl^- or Br^-). It is shown that the substitutions of F^- and Cl^- ions are not simply a mutual substitution, but really an interaction via hydrogen-bonding with the neighboring OH^- ions. This results in an additional binding with possible physiological significance.[11]

The structural location of the CO_3^{2-} ions has not yet decisively been revealed. What is known is that the substitution of CO_3^{2-} at the B site produces a contraction of the a-lattice parameter at a rate of 0.006 Å/wt% CO^3. It appears that the B CO_3^{2-} group is located at the oblique face of the PO_4 tetrahedron.[4] More information is available on the location of the A type CO_3^{2-} ions, which cause an expansion of the a-lattice parameter at a rate of 0.026 A/wt% CO_3. The CO_3^{2-} group is centered on the c-axis; its plane is tilted about 30° from the c-axis.[5] Many substitutions are still studied: generally spoken substitutions in the channels cause deformations in the crystal structure and influence the crystallization behavior of the apatites.[1]

III. SYNTHETIC VS. BIOLOGICAL APATITES

Although the above-described structure of hydroxyapatite has long been used as the idealized model for the structure of biological apatite, there are significant differences in crystallography and composition. Similarly, structural and compositional difference may also arise between the ideal apatite and actual synthetic hydroxyapatites. These differences can result from processing and manufacturing. In order to better understand the changes in

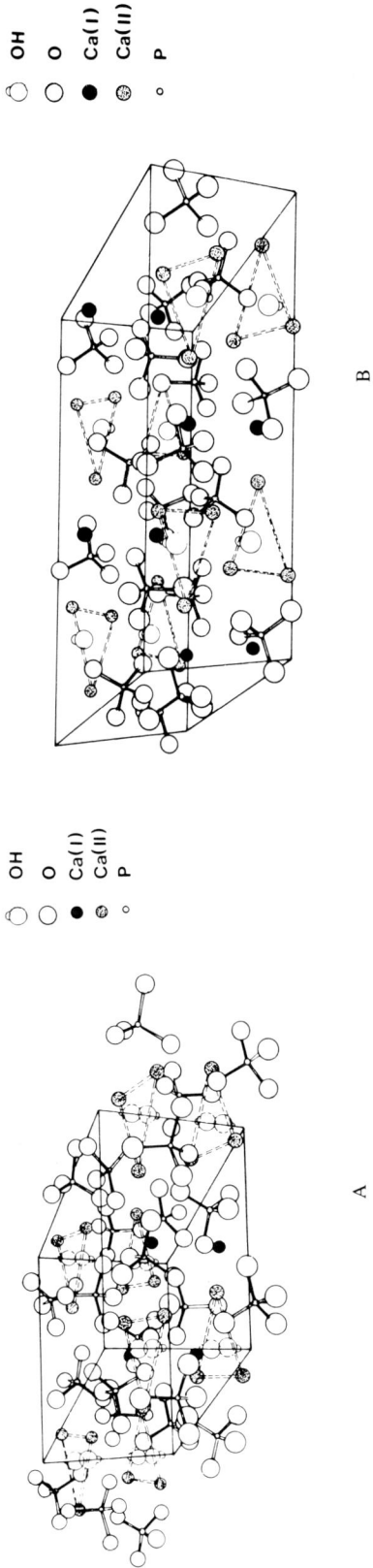

FIGURE 3. (A) Hexagonal structure of hydroxyapatite. The symbol ⊖ stands for the OH⁻ positions statistically 50% occupied; (B) monoclinic structure of hydroxyapatite.[11]

Table 1
MINERAL CONSTITUTION OF HARD TISSUES

| Element | Dental tissues[15] | | | Bone[16] |
	Enamel	Dentine	Cement	
Calcium	36.1	35	35.5	26.7
Phosphorus	17.3	17.1	17.1	12.47 (expressed as PO_4^{3-})
CO_2	3.0	4.0	4.4	3.48 (expressed as CO_3^{2-})
Mg	0.5	1.2	0.9	0.436
Na	0.2	0.2	1.1	0.731
K	0.3	0.07	0.1	0.055
Cl	0.3	0.03	0.1	0.08
F	0.016	0.017	0.015	0.07
S	0.1	0.2	0.6	—
Cu	0.01	—	—	—
Si	0.003	—	0.04	—
Fe	0.0025	—	0.09	—
Zn	0.016	0.018	—	—

structure and properties of these apatites, it is clarifying to consider some structure-property relationships of biological apatites.

Stoichiometric hydroxyapatites contain 39.9 wt% Ca, 18.5% P, and 3.38% OH: the atomic ratio Ca/P is 1,67; the mineral constituent of hard tissues contains not only Ca, PO_4, and OH ions, but also small amounts of CO_3, Mg, Na, or trace amounts of elements such as F and Cl. The amount of these ions for various hard tissues such as dentine, enamel, and cement of teeth or osseous tissue is shown in Table 1. In addition to the presence of these substituting ions, the Ca/P ratio differs from the stoichiometric value.

A first example of the effect of these structural variations pertains to the CO_3 content. The higher the percent CO_3 incorporated into biological apatites, the higher is the metabolic activity e.g., the content is lower in enamel (a practically inert material) than in dentine (a more active tissue). In addition it has been shown that only 10% of the CO_3 ions of untreated tooth enamel occupy A-sites, thus, the B-site is of greater biological relevance.[11]

Similarly there is an effect of the Ca/P ratio. The ratio increases from the most active to the least active zone. In this latter zone the Ca/P ratio is typically 1.66. The Ca/P ratio is smaller in younger than in older bones.[17-20]

Fluorine substitution in hydroxyapatite changes the monoclinic to a hexagonal structure and affects the diffusion along the Z ions channel, diminishing the reactivity and, presumably, the solubility of the apatites.[11,12,19,21-24]

As a result of the ionic substitutions, changes in the lattice parameters obviously occur. For example, the a-axis of human enamel apatite is about 0.02 Å larger than that of stoichiometric hydroxyapatite (9.44 vs. 9.42 Å). This same length of the a-axis can also be obtained by synthesizing carbonated apatites.[11,23] The effect of the degree of crystallinity on reactivity is comparable to the effect of CO_3 content and Ca/P ratio. The mineral constituent of the most active areas (dentine, bone tissue) is the least crystallized or may even be amorphous, while the mineral constituent of the least active zone (enamel) is better crystallized.[1,20,25]

Some insight in these interrelationships was obtained by physicochemical work with synthetic tricalciumphosphate, $Ca_3(PO_4)_2$.[25] Freshly precipitated tricalciumphosphate is amorphous, with an apatite-like structure and chemical formula, $Ca_9(PO_4)_6\square_2$. In the presence of water, the precipitated phosphate undergoes a transformation. An internal hydrolysis reaction (Equation 1) partially transforms the PO_4^{3-} ions into HPO_4^{2-} ions and produces a partial filling up of the Z-site vacancies by OH^- ions.

$$PO_4^{3-} + H_2O \rightarrow HPO_4^{2-} \quad + OH^- \tag{1}$$

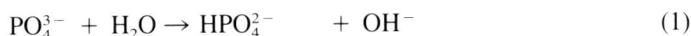

This hydrolysis reaction is endothermal and takes 24 hr at 20°C. The resulting product has the following chemical formula:

$$Ca_9\square \, (PO_4)_{6-x} \, (HPO_4)_x \, (OH)_x\square_{2-x}$$

Concurrent with the hydrolysis, a crystallization, resulting in an apatite structure, takes place. The hydrolysis reaction controls the crystallization: the amorphous structure is maintained when the hydrolysis reaction is inhibited. It was shown that various ions can delay the hydrolysis and thus the crystallization to an apatite. The presence of Mg^{2+} ions slows down the internal hydrolysis reaction due to the formation of $(Mg\,(H_2O)_6)^{2+}$, which hampers the crystallization. A comparable but less pronounced effect is found in the presence of $P_2O_7^{4-}$ and CO_3^{2-} ions. The simultaneous presence of these ions can cause a synergistic effect.[21]

Apart from differences in ionic species and crystallographic aspects, variations in specific surface can occur. Particularly noteworthy is the significant difference between bone mineral (100 to 200 m^2/g), synthetic hydroxyapatite (25 to 200 m^2/g), and synthetic amorphous calcium phosphate (20 to 60 m^2/g).[6,26]

IV. THERMAL BEHAVIOR

The manufacturing of calcium phosphates for biomedical applications, such as those that will be discussed in later sections, can include heat treatments. As such, knowledge about the thermal behavior is important. At high temperatures, the hydroxyapatite structure may be modified depending upon its stoichiometry, the temperature, the atmosphere, and the synthesis conditions.[7,27-32] In this paragraph a distinction is made between stoichiometric hydroxyapatites and apatites with Ca/P ratios different from 1.67.

A. Stoichiometric Apatites

Results from thermogravimetric, X-ray diffraction, and IR absorption analyses show the presence of two types of water in precipitated apatites.[31] These are adsorbed water and lattice water. Adsorbed or "loosely bound" water is responsible for a reversible weight loss without an effect on lattice parameters at temperatures from 25 to 200°C. Lattice water is irreversibly lost between 200 and 400°C. This causes a contraction in the a-lattice dimension. Possible sources of this lattice water are the presence of H_2O or HPO_4^{2-} which substitute OH^- or PO_4^{3-} ions, respectively, in the apatute lattice. The lattice water is not present in synthetic apatites prepared from nonaqueous systems, such as by solid state diffusion or by precipitation from a melt at temperatures between 900 and 1000°C.

At higher temperatures, the hydroxyapatite $Ca_{10}(PO_4)_6(OH)_2$ can be totally or partially dehydrated. Above 850°C a small weight loss is recorded; during cooling an equivalent increase in weight is observed. These observations indicate a slight reversible dehydration of the hydroxyapatite according to Reaction 2.

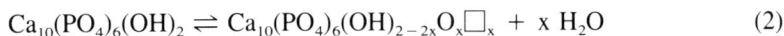

$$Ca_{10}(PO_4)_6(OH)_2 \rightleftharpoons Ca_{10}(PO_4)_6(OH)_{2-2x}O_x\square_x + x\, H_2O \tag{2}$$

This reaction is an equilibrium reaction and the reaction product may be influenced by the partial H_2O pressure. If H_2O is added, the reaction shifts to the left, thus stabilizing the hydroxyapatite. The structure of the hydroxyapatite is then preserved up to 1100°C.

Pyrolysis (decomposition as a result of heat) in vacuum or in gases free of water shifts the reaction to the right and an oxyhydroxyapatite is formed.

$$Ca_{10}(PO_4)_6(OH)_{2-2x}O_x\square_x$$

In the limit, when $x = 1$, oxyapatite, $Ca_{10}(PO_4)_6O$, is formed.[7] This reaction product exists only in the temperature range of 850 to 1050°C. It is highly reactive and at temperatures lower than 800°C, it undergoes a small rehydration giving rise to oxyhydroxyapatite. This is stable in air at room temperatures.[22-24]

From 1050°C on, the hydroxyapatite may decompose according to reaction 3.

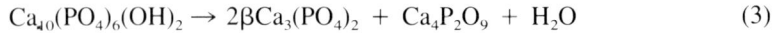

$$Ca_{10}(PO_4)_6(OH)_2 \rightarrow 2\beta Ca_3(PO_4)_2 + Ca_4P_2O_9 + H_2O \tag{3}$$

At temperatures higher than 1350°C, $\beta Ca_3(PO_4)_2$ transforms to $\alpha Ca_3(PO_4)_2$. The α phase, a polymorphic form of β, is the high-temperature phase which is maintained upon cooling.

It must be pointed out though that Skinner[29] finds that the decomposition of hydroxyapatite to $Ca_4P_2O_9$ and $Ca_3(PO_4)_2$ already starts at 850°C in vacuum. These results are not in agreement with the results of Trombe.[7-9]

Synthesis conditions can influence the thermal behavior of pure hydroxyapatite.[27,31] Newesely[27] compared the behavior of a precipitated apatite with a hydrothermal pure hydroxyapatite. In the former, the Ca/P ratio appears reduced because of the presence of HPO_4^{2-} ions. Decomposition of this hydroxyapatite is promoted and occurs at a lower temperature than the decomposition of the hydrothermally made hydroxapatite.

B. Nonstoichiometric Apatites

Nonstoichiometric hydroxyapatite powder is often used as starting material. The high-temperature behavior of these apatites differs from that of pure hydroxyapatites. In order to discuss the behavior of the hydroxyapatites, it is of value to consider the behavior of tricalciumphosphates.

Studies of the thermal behavior of precipitated tricalcium phosphates using X-ray diffraction, IR spectroscopy, thermogravimetric analyses (TGA), and differential thermal analysis (DTA) show that phosphates preserve their normal apatitic structure until 650°C without forming a supplementary phase.[25] Above 650°C tricalciumphosphate originates, and the reaction ends at 750°C. When heating up to 650°C, the externally and internally adsorbed water is lost. Reaction 4 takes place without a crystallographic change.

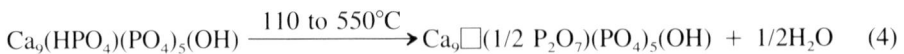

$$Ca_9(HPO_4)(PO_4)_5(OH) \xrightarrow{\text{110 to 550°C}} Ca_9\square(1/2\ P_2O_7)(PO_4)_5(OH) + 1/2H_2O \tag{4}$$

Reaction 4 is brought about by Reaction 5.

$$2HPO_4 \rightarrow P_2O_7^{4-} + H_2O \nearrow \tag{5}$$

Between 650 to 750°C, Reaction 6 takes place in the crystal, together with a crystallographic change giving rise to $\beta Ca_9(PO_4)_6$.

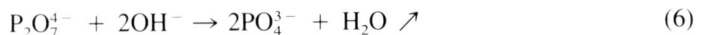

$$P_2O_7^{4-} + 2OH^- \rightarrow 2PO_4^{3-} + H_2O \nearrow \tag{6}$$

A generalization of these observations can be made when using commercial apatite products that are very often nonstoichiometric. A calcium-deficient hydroxyapatite can be represented by the chemical formula:

$$Ca_{10-x}(HPO_4)_x(PO_4)_{6-x}(OH)_{2-x}$$

Similar reactions to Reactions 4 and 5 can take place between 110 and 650°C. This gives rise to

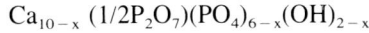

$$Ca_{10-x}(1/2P_2O_7)(PO_4)_{6-x}(OH)_{2-x}$$

On further heating to 700°C, Reaction 6 produces

$$Ca_{10-x}(x\ PO_4)(PO_4)_{6-x}(OH)_{2-8x/2}$$

This theoretical end product, however, does not exist, and a mixture of two products is obtained:

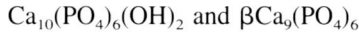

$$Ca_{10}(PO_4)_6(OH)_2\ \text{and}\ \beta Ca_9(PO_4)_6$$

The total reaction is written in Equation 7.

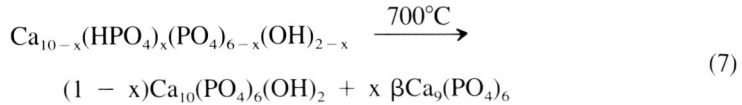

$$Ca_{10-x}(HPO_4)_x(PO_4)_{6-x}(OH)_{2-x} \xrightarrow{\ 700°C\ }$$

$$(1-x)Ca_{10}(PO_4)_6(OH)_2 + x\ \beta Ca_9(PO_4)_6 \qquad (7)$$

These reactions can be schematically summarized.

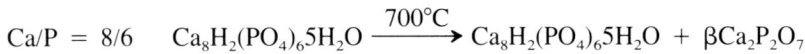

$$Ca/P = 10/6 \quad Ca_{10}(PO_4)_6(OH)_2 \xrightarrow{\ 1100°C\ } Ca_{10}(PO_4)_6(OH)_2$$

$$Ca/P = 10/9 \quad Ca_3(PO_4)_2 \xrightarrow{\ 700°C\ } Ca_3(PO_4)_2$$

$$Ca/P = 8/6 \quad Ca_8H_2(PO_4)_6 5H_2O \xrightarrow{\ 700°C\ } Ca_8H_2(PO_4)_6 5H_2O + \beta Ca_2P_2O_7$$

Calcium phosphates with intermediate Ca/P ratios produce mixtures of the respective products.[33-35]

V. SOLUBILITY

Many studies have discussed the solubility of hydroxyapatite in aqueous solutions. A large number of solubility constants have been reported. The variation is within the range 10^{-49} to 10^{-58}. Several mechanisms are proposed to explain the dissolution behavior of hydroxyapatite.[36-42]

1. Variations of solubility can be attributed to differences in the rate of formation and dissolution of an intermediate solid phase at the surface like $CaHPO_4 \cdot 2H_2O$ (brushite) and $CaHPO_4$(mo. etite).[36,39,42]
2. The solubility of hydroxyapatite is affected by the powder weight to liquid volume ratio. This can be explained by the ion exchanges that take place when a solid is submerged in a liquid. During equilibration experiments, dissolution and reprecipitation can cause an enrichment or loss of ions that change the solubility. The change in composition of the surface layer which influences the measurement of the solubility constant can be minimized by increasing the solid to solution ratio.[19,38,40,41]
3. Solubility of hydroxyapatite increases with decreasing pH. The effect of the surface to solution ratio becomes less important with decreasing pH.[35]
4. There is also an effect of the specific surface area. Surface ions may be held relatively

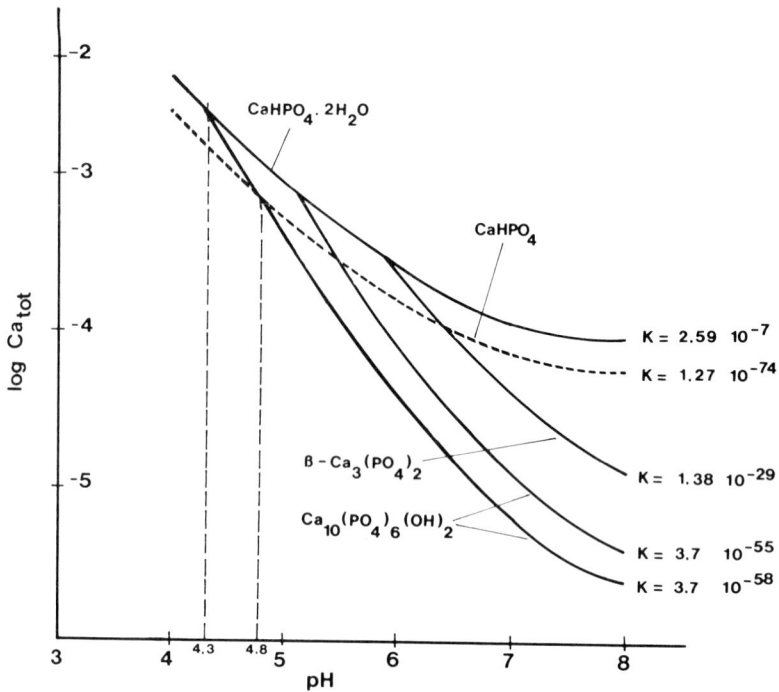

FIGURE 4.   Solubility isotherms for various phases in the system $CaO-P_2O_5-H_2O$ at 25°C.[11]

more weakly than the interior ions and may therefore be more soluble. Hydroxyapatite with a higher specific area has a greater solubility than the one with lower specific surface.[26,37]

5.   The highly polar surface of hydroxyapatite may result in an adherent film with a thickness of a few monolayers. The presence of this hydration sphere may influence the measurement of the ion concentration.[6]

6.   Crystal defects, such as impurities or vacancies may influence the dissolution phenomena by altering the exchange reactions in the diffusion controlling Z ion column.[43,44] There may also be an effect of dislocations on the diffusion phenomena.[45,47] The chemical reactivity of solids varies inversely to their crystalline perfection and the crystal dimension.[43,38]

7.   Many anions lower the solubility of hydroxyapatite at a moderate pH. Substitution of the OH^- ions by F^- lowers the solubility. Incorporation of FPO_3^{2-} and P_2O^{4-} also lowers the solubility. The exchange of F^- ions for OH^- ions is limited to the surface at low temperatures since intracrystalline exchange is slow. Its effect is, however, pronounced. Even a small percentage of F^- ions greatly increases the acid resistance. Hydroxyapatite treated with FPO_4^{2-} is more resistant than untreated hydroxyapatite.[22] On the other hand, the incorporation of Na^+ and CO_3^{2-} ions strongly increases the solubility of biological apatites.[19]

To illustrate some of the above mechanisms, two examples are given herewith. Brown and co-workers[49] reported on the solubility of calcium phosphates and found hydroxyapatite to be the most stable calcium phosphate above a pH of 4.8 at 25°C. Below pH 4.8 monetite, $CaHPO_4$, is the least soluble.

In Figure 4 the solubility isotherms for various calcium phosphates are plotted. Though monetite is more stable, the formation of brushite, $CaHPO_4 \cdot 2H_2O$, is kinetically more favorable. The intersection point of the isotherm of hydroxyapatite and brushite lies at pH 4.3. However, if the solubility product of the apatite shifts to higher values, the intersection of the solubility isotherms of apatite and brushite will shift to higher pH values. These considerations are important with respect to the understanding of which phases may be present in as manufactured apatite.

The second example is that some of the mineral in enamel appears to be more soluble than the synthetically prepared hydroxyapatite. A possible explanation is the presence of impurities in enamel which may change the solubility. Impurities such as carbonate chloride and sodium increase its solubility. These ionic substitutions are indeed found in vivo. Baud and Very[50] proved by X-ray diffraction and chemical analysis that fluorine, chlorine, strontium, and carbonate can be incorporated into the hydroxyapatite lattice of bone and tooth mineral. The presence of an amorphous hydroxyapatite with a greater reactivity may also be a plausible reason.[49]

VI. FABRICATION OF CALCIUM PHOSPHATE CERAMICS

A. Preparation of the Powder

The starting material for an implant is mostly a powder which is compressed and sintered. Hydroxyapatite powders can be prepared from aqueous or nonaqueous systems.

Hydroxyapatite from aqueous systems are prepared by either precipitation or hydrolysis of brushite ($CaHPO_4 \cdot 2H_2O$) or tricalciumphosphate, $Ca_3(PO_4)_2$, at temperatures below 100°C.[21,33,34,51-57] In the precipitation reaction, solutions of $Ca(NO_3)_2 \cdot 4H_2O$ and $(NH_4)_2HPO_4$ are brought to pH 11 to 12 with concentrated NH_4OH. The phosphate solution is dropwise added to the stirred Ca solution.[51,52,55,57]

Another method is the addition of H_3PO_4 to $Ca(OH)_2$; the pH changes from 12.4 to 8.7, as the precipitation reaction proceeds.[56] Hydrolysis of $CaHPO_4$ is carried out in a NaOH solution of low concentration for several hours at 25 to 100°C or for 20 hr in boiling distilled water.[21] Hydrolysis of $Ca_3(PO_4)_2$ at temperatures in the range of 40 to 80°C at pH 3 to 7.5 results in the formation of a calcium-deficient hydroxyapatite.[34-35] As a result of these different production ways, variations in compositional and structural properties are obvious. Important parameters are

1. The flow rate of the reagents, which controls the crystal size of the apatites
2. The pH value at which the precipitation is carried out
3. The presence of impurities which can easily be incorporated in the structure

Apatites from nonaqueous systems can be prepared by solid state reactions at temperatures above 900°C.[5,58,59] However, since most biomaterial applications are with apatites prepared via the aqueous route, no further attention is given to the solid state route. Most of these nonaqueous apatites have the ideal hydroxyapaptite crystal structure without lattice contractions. In contrast, presence of HPO_4 and/or lattice H_2O is possible in aqueous apatites. This results in a contraction of the a-axis during heating at 200 to 400°C and in the formation of a mixture of hydroxyapatite and tricalciumphosphate after pyrolysis at 850°C.

B. Compaction and Sintering of Calcium Phosphates

Most fabrication processes use high-pressure powder compaction techniques to produce a so-called green compact, which is further densified by sintering at temperatures from 1100 to 1300°C. With appropriate conditions both dense and porous ceramics can be produced. An important consideration concerns the term porosity.[60] Microporosity, with pores in the order of micrometers, relates to spaces that exist because of insufficient sintering. Macroporosity means the existence of larger pores that are often voluntarily introduced to enhance

tissue ingrowth. Following these porosity definitions a distinction will be made between dense ceramics (containing no micropores) and microporous ceramics. In the section on animal experiments and clinical applications, a ceramic with large pores will be called macroporous. A material may thus be simultaneously dense and macroporous.

Isostatic compaction is the forming technique which is often used to produce dense calcium phosphate ceramics. This results in a greater density and higher compressive strength of unsintered compacts than after uniaxial compression.[55,61-66]

Another fabrication technique consists of the simultaneous use of pressure and heat: this is the so-called continuous hot pressing technique (CHP). Densification takes place at a lower temperature (900°C) without the formation of $Ca_3(PO_4)_2$. However, no significant improvement in mechanical properties is observed.[67-69] A special forming process involves the collection of freshly precipitated hydroxyapatite. Drying of the slurry and subsequent sintering results in an almost completely dense ceramic called durapatite.[55,70]

There are several methods to introduce macropores. One method creates pores by drilling holes in the fired ceramic body.[71,72] The most widely used technique consists of mixing the calcium phosphate powder with appropriately sized organic powders, like cellulose or naphtalene. These powders have a lower burning temperature than the sintering temperature.[73-75] The evaporation of the organic material leaves macropores which are preserved during sintering. A third method relies on the dissociation of hydrogen peroxide in a powder slurry.[76] A fourth method is the reproduction of the microstructure of a coral by hydrothermal exchange. In the so-called replamineform process hydrothermal exchange of carbonate with phosphate is realized in an aqueous phosphate medium at elevated temperatures. This results in a calcium phosphate replica of the coral structure.[77-80]

C. Mechanical Properties

Depending upon the preparation and sintering conditions, a wide range of mechanical properties is obtained. A comparison of the observed mechanical properties of the calcium phosphate ceramics with hard tissues is made in Table 2. The large scatter is due to the variations in sample preparation. Fabrication of the powder, prior cold compaction, the proper sintering conditions (temperature, time, and cooling conditions), and the surface preparation of the test samples influence the characteristic properties of the calcium phosphate ceramic like:

1. The coexistence of hydroxyapatite and β-tricalciumphosphate as a result of the non-stoichiometry of starting material. The presence of $\beta Ca_3(PO_4)_2$ lowers the fracture strength. Samples of hydroxyapatite fired at 1150°C, containing no second phase, showed higher fracture strengths (98 MNm^{-2}) as compared to those containing 40 wt% $\beta Ca_3(PO_4)_2 MNm^{-2}$).[55] In another study it was shown that there was a decreasing influence on compressive strength of the coexistence of monocalciumphosphate with di-, tri-, and tetracalciumphosphate.[72]
2. The mechanical properties decrease with increasing amount of macro– and micropores.[61,76]

Comparable to most ceramics, calcium phosphate implants have low tensile strength. This makes them hardly suitable for use in applications with tensile stressing.

The values in Table 2 are obtained on dry specimens, tested in air. Physiological environment may have a negative influence upon the mechanical properties. Some evidence for this is found in a report describing mechanical tests carried out in distilled water on dense hydroxyapatite ceramics.[85] Both flexural strength and fracture toughness decrease 25% respective to the values found in a dry atmosphere. The slow crack growth, already observed to a small extent in dry condition, becomes considerably greater in wet conditions.

Table 2
COMPRESSIVE AND TENSILE STRENGTHS OF HARD TISSUES AND CALCIUM PHOSPHATE CERAMICS

Ceramics/hard tissues	Compressive strength (MPa)	Tensile strength (MPa)	Ref.
Cortical bone	135—160	69—110	81
Dentine	295	51.7	82, 83
Enamel	270	70	67
	384	130	82, 83
Porous calcium	30—170	4.8	76, 84
phosphates	6.9—20	—	84
Dense calcium	917	78—196	55
phosphates	120	61—113	61, 65
	308—509	38—48	65, 67
	390—430	115	67, 85
	800	—	85

VII. ANIMAL EXPERIMENTS AND CLINICAL APPLICATIONS

Early studies were carried out with slurries of calcium phosphate powders. The working hypothesis of these studies was that a local release of calcium ions would enhance bone growth.[85-89] It was found that the powders implanted in bone tissue were resorbed and replaced by bone. Except in some cases with tricalciumphosphates, no acceleration of the healing rate was found. However, recent experiments with mixtures of synthetic hydroxyapatite and physiological saline in rabbit femurs indicate accelerated formation of new bone.[80] Due to lack of mechanical strength, these studies had little clinical relevance. Only one clinical study is reported.[91] In addition to the use of hydroxyapatites, many studies used calcium phosphate ceramics.

A recurrent observation with too many studies is the limited materials characterization. As such, the interpretation of the observed phenomena is often difficult. In the section that follows materials data are given when available.

A. Bulk Material
1. Macroporous Calcium Phosphate Implants

In a preliminary biocompatibility study, Köster et al.[72] implanted compact and macro-porous cylindrical calcium phosphate ceramics with various Ca/P ratios in dog tibia. They did not mention the sintering procedure nor the structural information on the microporosity. The compact compounds with Ca/P ratios of, respectively, 1, 1.5, and 2 were all biocompatible. The optimum ratio with regard to resorption was about 1.5. In the macroporous specimens, macropores with a diameter of 0.8 to 1 mm were mechanically created. The diameter of the interconnecting channels was about one third the pore diameter.[71,72,92] Healing took place and no fibrous tissue between bone and ceramic was observed after various periods of time up to 10 months. The authors suggested that the resorption rate of $Ca_3(PO_4)_2$ was dependent upon the stress conditions: less loaded tricalciumphosphate was resorbed more slowly than highly loaded. The amount of loading was assessed purely on the basis of the weight of the animal. In contrast, tetracalciumphosphate, $Ca_4P_2O_9$, with a Ca/P ratio of 2, did not resorb, even when it was highly loaded.

Cameron et al.[75,84] used a macroporous (36%) ceramic obtained from tricalciumphosphate powders. The large interconnecting macropores (100 to 300 μm) were reported to be uniformly distributed. No information was given about microporosity. The tissue reaction of dog femurs was benign: when implanted in cancellous bone the ceramic was slowly resorbed and only occasionally were giant cells present. The authors suggested that the resorption behavior may indicate a potential use as bone graft material. The same material was implanted in rabbit ears, and a rapid invasion of tissue in the macropores was reported.[93] Dissolution of the ceramic occurred slowly without the appearance of new bone.

Experiments of Lemons et al.[94,95] confirmed the possibility of using porous tricalcium-phosphate to provide a scaffold for tissue ingrowth: complete ingrowth of bone was observed after 12 weeks of implantation in rabbits. The resorption of tricalciumphosphate ceramics, used as bone grafts in various animal experiments, was also reported by other workers, but without giving structural specifications.[96-98]

Holmes[80] implanted in dog mandibula the hydroxyapatite replamineform implants prepared according to White et al.[77] The implants were biocompatible and bone regeneration was found. This increased with time. The regenerated bone was of a woven type at 2 months and changed to a lamellar type after 6 months. After 12 months implantation, a degradation of 29% of the implant could be observed.

Rejda et al.[76, 99-101] developed a degradable phosphate ceramic starting from a slurry of commercially available hydroxyapatite powder and hydrogen peroxide. Macro- and micro-porosity was controlled by variations of hydrogen peroxide concentration and sintering temperature and time, respectively. The ceramic sintered at 1250°C for 6 hr contained a small (10%) amount of $Ca_3(PO_4)_2$ and had macropores of 150 to 250 μm and micropores of 0.5 to 1.5 μm.

Compressive strength for various macroporosities varied between 30 and 170 MN/m^2.[76] Specimens with a macroporosity of 20% and a microporosity of 30% were biologically investigated.[99,100] After a few weeks of implantation in rabbit tibia, bone tissue was formed directly on the interface between ceramic and surrounding tissue, without any foreign body reaction. After 1 month, ingrowth was obvious and the ceramic was partially resorbed. After 6 months almost the whole ceramic body was replaced by new bone.[101] Different clinical trials were carried out with the same basic material, but with 10% microporosity and 20% macroporosity.[102]

With seven patients the calcium phosphate ceramic was implanted to enlarge resorbed lower jawbones. Less resorption was observed than with use of autologous bone. The same ceramic was also used to fill spaces left after mastoidectomy in 5 patients and to reconstruct in part the middle ear (13 patients). Another application was sought in joining vertebrae in patients with spinal instability. After 1 year a small amount of resorption was observed: the implant height decreased from 10 to 9 mm.

Nery et al.[103] implanted tricalciumphosphate ceramics with dimensions of 15 × 5 × 5 mm and a macroporosity of 50% in surgically produced periodontal osseous defects in dogs. The dimensions of the macropores varied between 800 and 1000 μ. Bone ingrowth occurred without toxic reactions. Recent experiments under functional loading with a unilateral partial denture with the same material with macropore diameters between 400 and 500 μm confirmed with earlier findings.[104] At the end of the 6-month implantation period, bone ingrowth was observed without inflammation and no bone resorption had occurred. A clinical study with a limited number of patients (six) indicated a good tolerance. After 1 year of implantation a slow resorption phenomenon might have been occurring.[105]

2. Dense Calcium Phosphate Implants

Jarcho et al.[55,106] implanted dense hydroxyapatite ceramics in dog femurs. Bone was found to grow upon the surface of the so-called durapatite implants. An intimate bond, probably due to a direct chemical attachment was formed. After 6 months no degradation of the implants was observed. The material was also used in dental applications. Kent et al.[107] evaluated the durapatite in the augmentation of deficient alveolar ridges in 30 patients. Radiographic examination after 18 months showed an excellent stability, and no adverse tissue reactions were found. Experiments with the durapatite root implants showed that the length of the plugs was not important, since with both long and short tooth plugs postextraction alveolar ridge resorption did not occur.[108]

Experiments by Aoki et al.[109] with dense implants under functional loading showed direct bone growth after 8 months: a low-grade type of inflammation of the gingival tissue was observed. Later on, it was reported that clinical trials were under way.[110]

Denissen et al.[67-69] used dense hydroxyapatite prepared in two ways: (1) isostatic compression and sintering of powder (densities of 97 and 99.9%, respectively, were obtained.) and (2) continuous hot pressing of powders, resulting in a density of 97%. Implantation of the differently prepared ceramic materials in rat tibia and dog alveolar bone gave excellent results. In both cases, new bone was laid down almost immediately next to the implant surface. Both in rat and dog experiments a thin amorphous layer between implant and surrounding tissue was observed. In the tibia implants, the amorphous layer amounted to 60 nm; in the alveolar experiments, the width of the amorphous layer varied between 0 and 1000 nm. Probably this layer plays an important role in the strong bonding that was observed between implant and living bone: it is conceivable that this bond is a chemical one. The dense ceramics appear fully compatible with the rat tibia, and after an implantation period of 6 months, no degradation was found. A total of 100 hydroxyapatite implants was used clinically as synthetic toothroots. The healing of the wounds occurred without complications. After 30 months neither rejection nor resorption were found and the implants seemed to be effective in maintaining the height and width of postextractive alveolar ridges.[68] Dense hydroxyapatite abutments were prepared in order to provide an interface material between cores and alveolar bone. After 1 year of implantation in 20 patients, it was observed on radiographs that alveolar bone closely surrounded the implant.[111] Dense hydroxyapatite blocks of greater dimensions were implanted in dog mandibles for 12 weeks. The model was approaching the above clinical situation. Direct bone deposition and firm attachment of the implants were observed.[112]

B. Other Developments

Calcium phosphate ceramics have low tensile properties. This limits their application in bulk form. To use these materials in tensile loaded conditions, the reinforcement of the calcium phosphate ceramics or the use as a coating on a suitable substrate has been studied. Rejda et al.[101] suggested that the sintered tricalciumphosphate would achieve a higher bending strength by filling the pores of the ceramic with a biologically degrading polymer. The latter possibility was already tried by Riess et al.,[113-116] who coated cylindrical implantable tooth roots of titanium with tricalciumphosphate, developed by Köster et al.[71,72,92] The implants consisted of successive layers of polymethylmethacrylate (PMMA) which was used as binder, the almost nondegradable tetracalciumphosphate,[72,92] and the degradable tricalciumphosphate. The thickness of the upper tricalciumphosphate layer was 2 to 3 mm. However, further studies are needed to create a direct coating, without use of the less compatible PMMA.[116] The present authors studied another form of calcium phosphate coating:[117] hydroxyapatite was deposited by depositing a thin lining on the inner pore walls of a stainless steel fiber structure. A stimulation of bone infiltration and formation within the pore structure was observed.

A new concept is the development of calcium phosphate containing bone cements: with the addition of apatite particles a direct chemical bond between bone cement and bone tissue may be possible.[118]

C. Ceramics with Ions, Additional to Calcium and Phosphate

Apart from calcium phosphates, some ceramics with ions additional to calcium and phosphate were investigated as potential implant material. Chiroff et al.[78] studied a $CaCO_3$ skeleton structure made by the replamineform process. After 1 year of implantation in dogs, direct apposition of new bone was observed without encapsulation, but a rather fast resorption had occurred.

Calcium sulfate, $CaSO_4$, has long been used as a space filler. It is very well accepted by living tissues, but it is soluble. The rate of degradation may be controlled by the use of low resorbable polymers.[119]

Sintering of an equimolar mixture of $Ca_3(PO_4)_2$ and $MgAl_2O_4$ at 1150°C resulted in a hard ceramic with a porosity of 1.2%, consisting of a $Ca_3(PO_4)_2$ and $MgAl_2O$ phase.[120] Tooth roots with controlled dense and porous regions were implanted in dogs. After 12 months a strong attachment was observed.

Two-phased calcium aluminates (a CaO.b Al_2O_3) with a density of 65 to 70% were implanted in monkey femurs for 1 year and were found compatible without toxic reaction.[121,122] Resorption of a slightly more soluble phase of the implant resulted in infiltration of new bone in the enlarged ceramic pores. An increase in rate of bone growth, however, was not shown. Hammer et al.[123] implanted calcium aluminates with 70% porosity, with interconnecting open pores of 150 to 200 μm, to augment deficient dog mandibular ridges. The ceramic was well tolerated and bone ingrowth was observed without inflammatory reactions. Resorption was not observed after 1 year. Alumino-calcium-phosphorus oxide ceramic (50% Al_2O_3, 37% CaO, 13% P_2O_5) used to reconstruct mandibular bone defects in rabbits, was found to be resorbable.[124]

All these experiments indicate that many ceramics with ions additional to calcium or phosphorus containing groups are biocompatible and show little adverse tissue reactions. However, in many cases the usefulness of these ceramics for bone growth stimulation is still to be proved. The resorption rate can be higher than the rate of new bone formation. Because of the good results obtained with ceramics purely consisting of calcium phosphate salts, use of ceramics with additional elements will only be attractive if they show superior in vivo characteristics.

VIII. DISCUSSION

Calcium phosphate ceramics obviously show attractive properties, such as lack of toxicity, absence of intervening fibrous tissue, the possibility of forming a direct bond with bone, and the possibility to stimulate bone growth. The effect of the surrounding tissues and fluids on the rate of biodegradation of the calcium phosphate ceramics, however, is less uniform.

A. Biodegradation

A wide range of interaction patterns was found. Resorption, partial resorption, or stability was observed. The way in which calcium phosphate implants undergo degradation is uncertain: it may be either a chemical dissolution or a cellular metabolic process. An understanding of this degradation behavior can be gained by considering both composition and material characteristics of the ceramics.

1. Chemical Composition

Most of the calcium phosphate ceramics implanted are made from starting powders prepared from aqueous solutions. As explained in an earlier section, both the purity of the starting material and the high-temperature treatment affect the crystal structure and the chemical composition of the ceramic. The presence of foreign elements like F, CO_3, or Mg in the ceramic structure may influence the behavior of the ceramic implant in a similar way as they do in natural hard tissues. Until now, this had not been investigated thoroughly, but some in vitro[125] and in vivo[126] experiments sustain this view. A comparison of various dense calcium phosphate ceramics implanted in rat femurs revealed different reactions. These results indicated that preparation conditions are important to assess the in vivo behavior. Definite explanations were, however, not given.

Most of the implant materials which are currently under investigation are composed of hydroxyapatite or tricalciumphosphate or a combination of the two. From fundamental studies it is known that the tricalciumphosphate powders dissolve more rapidly than hydroxyapatites. Although comparative in vitro studies on dissolution rates of both dense hydroxyapatite and tricalciumphosphate ceramics are not reported, it may be expected that the tricalciumphosphate ceramics dissolve faster than hydroxyapatite ceramics. This can be concluded from in vivo studies of porous calcium phosphates of similar structure, but with various compositions: tetracalciumphosphate implants degraded much slower than tricalciumphosphate implants.[92] In another comparative study, porous tricalciumphosphate ceramics degraded faster than porous hydroxyapatite implants.[126] Comparison of results of different research groups is hardly possible because one has to take into account that differences in degradation rate may also be due to the type and age of animal and to the localization and stress conditions of the implant.

2. Material Characteristics

Material characteristics are another group of parameters which control the resorption. Dissolution rates depend on the specific surface area. Due to their small specific surface area, densely sintered ceramics show a smaller tendency to degrade than porous ceramics. Little is known about the factors governing the resorption rate of different porous calcium phosphate ceramics. A comparison among the various studies is very difficult because many studies do not present sufficient materials characterization.

A factor in explaining biodegradability is the microporosity of the ceramic implant. De Groot[60] suggests that with identical internal surface the biodegradation of a ceramic implant, either tricalciumphosphate or hydroxyapatite, will be the same. It is indeed possible that micropores control the rate of biodegradation. The biodegradation process with successive release of individual particles and intracellular digestion is difficult for dense ceramics that will therefore show no degradation especially in the relatively short times applied in animal experimentation.

B. Inflammatory Reactions

The close chemical correspondence between calcium phosphate ceramics and natural hard tissues is a possible explanation for the observed disappearance of foreign body responses after prolonged periods of time. Analyses of calcium and phosphate levels of serum and urine during implantation studies of fast and slowly degradable calcium phosphate ceramics produced normal levels.[84,98,105,127] Long-term inflammation responses are sometimes reported. Kaiser et al.[128] reported fistula formation and rejection of tricalciumphosphate ceramics 10 weeks after implantation in bony defects of man. Winter et al.[126] also observed adverse tissue reactions of certain hydroxyapatite ceramics implanted in rat femurs. These inflammatory responses were probably caused by the release of impurities incorporated in the ceramic. Except in the two above cases, the histologic phenomena observed at the interface of dense implants with bone can be characterized as normal bone healing processes.[67-69,106,111-112] The increase in Ca/P ratio and of calcium and phosphorus content with implantation time indicates that normal calcification takes place.[106,109]

Initial stabilization of the implant was considered as a main prerequisite to achieve effective healing.[84,98,99,106,127] New bone was formed without intervening fibrous tissue. When a calcium phosphate ceramic implant was placed in a soft tissue environment, away from a bony structure, the implant surface was covered with fibrous tissue and no bone formation was observed.[84,126,127,129]

C. Tissue Bonding

A tissue bonding is observed which does not depend on tissue ingrowth. When an implant is removed from the surrounding tissue, separation rarely occurs at the interface.[67,106] The exact chemical composition of the reaction layer between implant and bone is not known. Electron microscopic studies would indicate that the narrow bonding zone shrinks with time in width from a few micrometers to 500 to 2000 Å.[67,106]

Jarcho et al.[106] who observed a perpendicular attachment of bone at the ceramic interface with a change to a parallel orientation at a distance of 100 Å from the interface, suggested a direct epitaxial deposition. However, recent results[110,130] indicate that the bonding layer may vary within single implant specimens. The authors suggested that a final identification of the bonding interface depends on a full characterization of the material.

D. The Stimulation of Bone Growth

The bone growth stimulating effect of calcium phosphate implants in bony structures received considerable attention. Early experiments already led to controversial results. Some investigators observed an enhanced formation of new bone,[86,89] while others reported no increase in bone growth rates.[87,88] Later, various experiments were set up to compare the behavior of calcium phosphate implants with autogenous material and empty defects as controls.

In almost all cases, autogenous material overrides the calcium phosphate ceramics with regard to bone healing rates. Levin et al.[74] compared tricalciumphosphate with cancellous marrow in the treatment of periodontal defects in dogs. The ceramic areas showed a slower healing rate. Shima et al.[131] implanted the porous Synthos material in cervical intervertebral spaces in dogs and found that "new bone formation was never as extensive as in autologous bone graft sites."

In millipore chambers implanted in dog femurs, the control chambers containing marrow alone contained more bone matrix than those filled with marrow and hydroxyapatite ceramic particles.[132] Lemons et al.[95] studied the bone bridging capacity in rabbit tibia and dog radii lesions of porous tricalciumphosphate ceramic and autogenous bone. The autogenous material showed superior bridging rates. However, a research group led by Köster[133] reported that their tricalciumphosphate ceramic showed a comparable osteogenic effect as autologous spongiosa, when implanted in dog tibia.

When tricalciumphosphate ceramics were compared with empty control defects, both slower and faster healing rates were reported. Nery et al.[103] and Levin et al.[74] reported a faster healing of empty spaces in periodontal defects in dogs than those filled with porous tricalciumphosphate ceramics. In contrast to these results, Mors and Kaminsky[98] found the healing process of surgically created defects in dog palates to be faster in the defect sites filled with tricalciumphosphate ceramic than in the empty control sites. The same results were reported by Ferraro[127] using porous tricalciumphosphate ceramics as bone grafts in orbital rims, mandibles, and iliac crests of dogs. Osborn and Newesely[134] recently observed osteogenic properties of calcium phosphate ceramics. Another evidence of the osteogenic properties of hydroxyapatite was given by experiments of the present authors.[117] A thin layer of hydroxyapatite on a metal fiber structure stimulated bone infiltration within the pores.

Although calcium phosphate ceramics are probably not as good bone inducers as autogenous material, they do enhance bone growth. The way in which calcium phosphate ceramics increase bone growth is not completely understood. Possible explanations are furnished by crystal growth mechanisms; either precipitation from a supersaturated solution or epitaxial growth. A slow resorption of a degradable ceramic provides excess calcium and phosphate that further crystallizes towards hydroxyapatite.[19,135] Another possibility, more likely applicable in almost nondegradable ceramics is the epitaxial growth of new crystals upon the implant surface.[106,134]

E. Limitations and Possible Applications

As evidenced by the foregoing paragraph both porous and dense calcium phosphate ceramics show promising applications. In various animal studies, porous, thus more or less degrading, calcium phosphate ceramics were used as bone grafts with the aim to serve as a scaffold for repair. Some clinical trials for joining vertebrae and in middle ear reconstruction are underway.[102]

Dense calcium phosphate ceramics have mechanical properties comparable to natural hard tissues. Animal experiments indicate that these dense implants are useful in sites mainly subjected to compressive stresses. In clinical trials these dense ceramics have been used for enlargement of jawbones[107] and as root implants to prevent postextraction alveolar ridge resorption.[67,68,108,110,111]

Despite many promising results both in animals and in humans, it is obvious that still significant problems have to be solved. As already evidenced by an in vitro study,[85] physiological attack lowers the mechanical properties. The same phenomenon was observed in recent experiments with transmucosal dense hydroxyapatites in dogs.[136] After a number of months of functional use, the ceramic fractured.

Because of the limited fatigue properties, long-term functionality of tension-loaded bulk calcium phosphate ceramics is doubtful. This long-term functionality is an essential condition for a biomaterial. Cranin[137] states that a dental implant must provide 5 years of satisfactory function to be judged successful. With the higher bioactivity of animal tissues in mind, the implantation times currently attained in patients are essentially too short to judge the clinical behavior. Further long-time animal and clinical studies are needed to assess the role of dense calcium phosphate implants. An alternative use of calcium phosphate ceramics might be the application of calcium phosphate coatings.[113-116] A tempting idea is the deposition of hydroxyapatite linings onto porous metal structures. The combination of the bone growth stimulation properties of the surface active calcium phosphate with the excellent mechanical properties of the metal may result in an efficient bonding.[117]

IX. CONCLUSIONS

Animal experimentation and limited clinical applications confirm the usefulness of calcium phosphate ceramics. Considering the numerous experimental results with a variety of implant-tissue interface reactions, in vivo behavior remains a complex phenomenon which is not yet completely understood. In a number of studies, lack of exact data on material properties inhibit a thorough evaluation. It is obvious that the explanation of the in vivo behavior of the calcium phosphate ceramics necessitates a complete materials characterization. Besides information about the specific clinical situation like the metabolic activity of the tissue, complete knowledge about the processing of the bioceramics and their final chemical, physical, and crystallographic features is necessary in order to assess local and systemic consequences of the implantation.

With increasing nondegradability, specific promising clinical applications, for example, are in the use as bone grafts, augmentation of alveolar ridges, and implantable teeth. Tensile stressed clinical applications of calcium phosphate ceramics in bulk form will probably not be possible. Methods to use calcium phosphate ceramic under these conditions have still to be worked out.

REFERENCES

1. **Montel, G.,** Conceptions actuelles sur la structure et la constitution des apatites synthétiques comparables aux apatites biologiques, in *Colloques internationaux du CNRS. Physico-chemie et crystallographie des Apatites d'intérêt biologique,* CNRS, Paris, 1975, 13.

2. **Bonel, G. and Montel, G.,** Sur l'introduction des ions CO_3^{2-} dans le réseau des apatites calciques, *C. R. Acad. Sci.,* 263. 1010, 1966.

3. **Trombe, J. C., Bonel, G., and Montel, G.,** Sur les apatites carbonatées préparées à haute température, *Bull. Soc. Chim. Fr.,* special issue, 1708, 1968.

4. **Legeros, R. Z., Trautz, O. R., Legeros, J. P., and Klein, E.,** Carbonate substitution in the apatite structure (1), *Bull Soc. Chim. Fr.,* special issue, 1712, 1968.

5. **Bonel, G.,** Contribution à l'étude de la carbonation des apatites, *Ann. Chim. (Paris),* 7, 65, 1972.

6. **Kibby, C. L. and Hall, W.,** Surface properties of calciumphosphates, in *The Chemistry of Biosurfaces,* Hair, M. L., Ed., Marcel Dekker, New York, 1972, chap. 15.

7. **Trombe, J. C.,** Contribution à l'étude de la décomposition et de la Réactivité de Certaines Apatites Hydroxylées et Carbonatées ou Fluorées Alcaline-Terreuses, Ph.D. thesis, L'Institute National Polytechnique de Toulouse, Toulouse, 1972.

8. **Trombe, J. C. and Montel, G.,** Some features of the incorporation of oxygen in different oxidation states in the apatite lattice. I. On the existence of calcium and strontium oxyapatites, *J. Inorg. Nucl. Chem.,* 40, 15, 1978.

9. **Trombe, J. C. and Montel, G.,** Some features of the incorporation of oxygen in different oxidation states in the apatitic lattice. II. On the synthesis and properties of calcium and strontium peroxyapatite, *J. Inorg. Nucl. Chem.,* 40, 23, 1978.

10. **Rey, C., Trombe, J. C., and Montel, G.,** Some features of the incorporation of oxygen in different oxidation states in the apatitic lattice. III. Synthesis and properties of some oxygenated apatites, *J. Inorg. Nucl. Chem.,* 40, 27, 1978.

11. **Young, R. A.,** Some aspects of crystal structure modeling of biological apatites, in *colloques Internationaux du CNRS. Physico-chimie et cristallographie des Apatites d'Intérêt Biologique,* CNRS, Paris, 1975, 21.

12. **Elliott, J. C. and Mackie, P. E.,** Monoclinic hydroxyapatite, in *Colloques internationaux du CNRS. Physico-chimie et Cristallographie des Apatites d'Intérêt biologique,* CNRS, Paris, 69.

13. **Kay, M. I., Young, R. A., and Posner, A. S.,** Crystal structure of hydroxyapatite, *Nature (London),* 204, 1050, 1964.

14. **Elliott, J. C., Mackie, P. E., and Young, R. A.,** Monoclinic hydroxyapatite, *Science,* 180, 1055, 1973.

15. **Sicher, H.,** *Orban's Oral Histology and Embriology,* 5th ed., C.V. Mosby, St. Louis, 1962, 53.

16. **Jaffe, H. L.,** *Metabolic, Degenerative and Inflammatory Diseases of Bones and Joints,* Lea & Febiger, Philadelphia, 1972, 125.

17. **Münzenberg, K. J.,** Untersuchungen zur Kristallographie der Knochenminerale, *Biomineralisation,* 1, 67, 1970.

18. **Münzenberg, K. J. and Gebhardt, M.,** Brushite, Octacalciumphosphate and carbonate-containing apatite in bone, *Clin. Orthop. Relat. Res.,* 90, 271, 1973.

19. **Driessens, F. C. M., Van Dijk, J. W. E., and Borggreven, J. M. P. M.,** Biological calcium phosphates and their role in the physiology of bone and dental tissues. I. Composition and solubility of calcium phosphates, *Calcif. Tissue Res.,* 26, 127, 1978.

20. **Glick, P. L.,** Ultrastructural aspects of dentin mineralization, paper presented at the 26th Annual Meeting of the ORS, Atlanta, February 5, 1980, 26.

21. **Legeros, R. Z., Shirra, W. P., Miravite, M., and Legeros, J. P.** Amorphous calcium phosphates synthetic and biological, in *Colloques Internationaux du CNRS. Physico-chimie et Cristallographie des Apatites d'Intérêt Biologique,* CNRS, Paris, 1975, 105.

22. **Ingram, G. S.,** Some heteroanionic exchange reactions of hydroxyapatite, *Bull. Soc. Chim. Fr.,* special issue, 1841, 1968.

23. **Young, R. A.,** Biological apatite versus hydroxyapatite at the atomic level, *Clin. Orthop.,* 113, 249, 1975.

24. **Brown, W. E. and Chow, L. C.,** Chemical properties of bone mineral, in *Annual Review of Materials Science,* Huggins, R. A., Bube, R. H., and Robert, K. W., Eds., Palo Alto, Calif., 1976, 213.

25. **Heughebaert, J. C.,** Contribution à l'Étude de l'Évolution des Orthophosphate de Calcium Précipités Amorphes et Orthophosphates Apatitiques, Ph.D. thesis, Institut National Polytechnique de Toulouse, Toulouse, 1977.

26. **Posner, A. S. and Beebe, R. A.,** The surface chemistry of bone mineral and related calcium phosphates, *Semin. Arthritis Rheum.,* 4, 267, 1975.

27. **Newesely, H.,** High temperature behavior of hydroxyapatite and fluorapatite, *J. Oral Rehabil.,* 4, 97, 1977.

28. **Kutty, T. R. N.,** Thermal decomposition of hydroxyapatite, *Indian J. Cem.,* 11, 695, 1973.

29. **Skinner, H. C. W., Kittelberger, J. S., and Beebe, R. A.,** Thermal instability in synthetic hydroxyapatite, *J. Phys. Chem.,* 74, 2017, 1975.

30. **Montel, G.,** Constitutions et structure des apatites biologiques: influence de ces facteurs sur leurs propriétés, *Biol. Cell.,* 2, 197, 1977.
31. **Legeros, R. Z., Bonel, G., and Legros, R.,** Types of "H₂O" in human enamel and in precipated apatites, *Calcif. Tissue Res.,* 26, 111, 1978.
32. **Füredi-Milhofer, H., Hlady, V., Baker, F. C., Beebe, R. A., Wikholm, N. W., and Kittelberger, J. S.,** Temperature programmed dehydration of hydroxyapatite, *J. Colloid Interface Sci.,* 70, 1, 1979.
33. **Berry, E. E.,** The structure and composition of some calcium deficient apatites, *Bull. Soc. Chim. Fr.,* special issue, 1765, 1968.
34. **Momna, H.,** Preparation of octacalcium phosphate by the hydrolysis of tricalcium phosphate, *J. Mater. Sci.,* 15, 2428, 1980.
35. **Momna, H., Veno, S., and Kanazawa, T.,** Properties of hydroxyapatite prepared by the hydrolysis of tricalcium phosphate, *J. Chem. Tech. Biotechnol.,* 31, 15, 1981.
36. **Rootare, H. M., Deitz, V. R., and Carpenter, F. G.,** Solubility product phenomena in hydroxyapatite-water systems, *J. Colloid Interface Sci.,* 17, 179, 1962.
37. **Golutvina, M. M., Semochkina, L. S., and Ilyushchenko, O. N.,** Determination of the specific surface of bone and synthetic hydroxyapatite, *Radiokhim. Acta,* 14, 502, 1972.
38. **Chuong, R.,** Experimental study of surface and lattice effect on the solubility of hydroxyapatite, *J. Dent. Res.,* 52, 911, 1973.
39. **Narasarju, T. S. B. and Rao, V. L. N.,** A new interpretation of the solubility equilibria of hydroxyapatite, *Z. Phys. Chem. (Leipzig),* 255, 655, 1974.
40. **Mika, H., Bell, L. C., and Kruger, B. J.,** The role of surface reactions in the dissolution of stoichiometric hydroxyapatite, *Arch. Oral Biol.,* 21, 697, 1976.
41. **Bell, L. C., Mika, H., and Kruger, B. J.,** Synthetic hydroxyapatite solubility product and stoichiometry of dissolution, *Arch. Oral Biol.,* 23, 329, 1978.
42. **Kaufman, H. W. and Kleinberg, I.,** Studies on the incongruent solubility of hydroxyapatite, *Calcif. Tissue Int.,* 27, 143, 1979.
43. **Montel, G.,** Rôle de la recherche fondamentale dans la conception des biocéramiques, paper presented at Cours de formation et de recyclage en céramique, University of Mons, October 8, 1979, IV.5.
44. **Langdon, B., Dykes, E., and Fearnhead, R. W.,** Defects, diffusion and dissolution in biological and synthetic apatite, in *Colloque internationaux du CNRS. Physico-chimie et Cristallographie des Apatites d'Intérêt Biologique,* CNRS, Paris, 1975, 381.
45. **Arends, J., Van Den Berg, P. J., and Jongebloed, W. L.,** Dissolution of hydroxyapatite and fluorapatite single crystals, in *Colloques Internationaux du CNRS. Physico-chimie et Cristallographie des apatites d'Intérêt Biologique,* CNRS, Paris, 1975, 389.
46. **Arends, J. and Jongebloed, W. L.,** Mechanism of enamel dissolution and its prevention, *J. Biol. Buccale,* 5, 219, 1977.
47. **Arends, J. and Jongebloed, W. L.,** Ultrastructural studies of synthetic apatite crystals, *J. Dent. Res.,* special issue B, 837, 1979.
48. **Blumenthal, N. C., Betts, F., and Posner, A. S.,** Formation and biological significance of calcium-deficient hydroxyapatites, paper presented at the 25th Annual Meeting of the ORS, San Francisco, February 20, 1979, 119.
49. **Brown, W. E.,** Physicochemistry of apatite dissolution, in *Colloques Internationaux du CNRS. Physico-chimie et Cristallographie des Apatites d'Intérêt biologique,* CNRS, Paris, 1975, 355.
50. **Baud, C. A. and Very, J. M.,** Ionic substitutions, in vivo, in bone and tooth apatite crystals, in *Colloques Internationaux du CNRS. Physico-chimie et Cristallographie des Apatites d'Intérêt Biologique,* CNRS, Paris, 1975, 405.
51. **Hayek, E. and Newesely, H.,** Pentacalciummonohydroxyorthophosphate, *Inorg. Synth.,* 7, 63, 1963.
52. **Moreno, E. C., Gregory, T. M., and Brown, W. E.,** Preparation and solubility of hydroxyapatite, *Natl. Bur. Stand.,* 72a, 773, 1968.
53. **Smith, A. N., Posner, A. M., and Quirk, J. P.,** Incongruent dissolution and surface complexes of hydroxyapatite, *J. Colloid Interface Sci.,* 48, 442, 1974.
54. **Lerch, P., Delay, A., and Friedli, C.,** Etude de la composition chimique et de la structure de composés de synthéses semblables aux apatites biologiques. Mécanisme d'élévation du rapport molaire Ca/P, in *Colloques internationaux du CNRS. Physico-chimie et Cristallographie des Apatites d'Intérêt Biologique,* CNRS, Paris, 1975, 85.
55. **Jarcho, M., Bolen, C. H., Thomas, M. B., Bobick, J., Kay, J. P., and Doremus, R. H.,** Hydroxylapatite synthesis and characterization in dense polycrystalline form, *J. Mater. Sci.,* 11, 2027, 1976.
56. **Tagai, H. and Aoki, H.,** Preparation of synthetic hydroxyapatite and sintering of apatite ceramic, paper presented at the Bioceramics Symposium, Keele, September 16, 1978.
57. **Arends, J., Schuthof, J., Vanderlinden, W. H., Bennema, P., and Van den Berg, P. J.,** Preparation of pure hydroxyapatite single crystals by hydrothermal recrystallization, *J. Cryst. Growth,* 46, 213, 1979.

58. **Elliott, J. C. and Young, R. A.,** Conversion of single crystals of chlorapatites into single crystals of hydroxyapatites, *Nature (London)*, 214, 904, 1967.
59. **Prener, J. S.,** The growth and crystallographic properties of calcium, fluor and chlorapatite crystals, *J. Electrochem. Soc.*, 114, 77, 1967.
60. **De Groot, K.,** Bioceramics consisting of calcium phosphate salts, *Biomaterials*, 1, 47, 1980.
61. **Rao, R. W. R. and Boehm, R. F.,** A study of sintered apatite, *J. Dent. Res.*, 53, 1351, 1974.
62. **Osborn, J. F. and Newesely, H.,** The material science of calcium phosphate ceramics, *Biomaterials*, 1, 108, 1980.
63. **Rootare, H. M. and Craig, R. G.,** Characterization of hydroxyapatite powders and compacts at room temperature and after sintering at 1200°C, *J. Oral Rehabil.*, 5, 293, 1978.
64. **Rootare, H. M. and Craig, R. G.,** Characterization of the compaction and sintering of hydroxyapatite powders by mercury porosimetry, *Powder Technol.*, 9, 199, 1974.
65. **Akao, M., Aoki, H., and Kato, K.,** Mechanical properties of sintered hydroxyapatite for prosthetic application, *J. Mater. Sci.*, 16, 809, 1981.
66. **Bigi, A., Incerti, A., Roveri, N., Foresti-Serantoni, E., Mongiorgi, R., Riwa di Sanseverino, L., Krajewski, A., and Ravaglioli, A.,** Characterization of synthetic apatites for bioceramic implants, *Biomaterials*, 1, 140, 1980.
67. **Denissen, H. W.,** Dental Root Implants of Apatite Ceramics, Ph.D. thesis, Free University of Amsterdam, Amsterdam, 1977.
68. **Denissen, H. W., de Groot, K., Makkes, P. Ch., Van den Hooff, A., and Klopper, P. J.,** Animal and human studies of sintered hydroxylapatite as a material for tooth root implants, Paper 3.8.1, presented at the 1st World Biomaterials Congress, Baden near Vienna, April 8, 1980.
69. **Denissen, H. W., de Groot, K., Makkes, P.Ch., Van den Hooff, A., and Klopper, P. J.,** Tissue response to dense apatites in rats, *J. Biomed. Mater. Res.*, 14, 713, 1980.
70. **Jarcho, M., Salsbury, R. L., Thomas, M. B., and Doremus, R. H.,** Synthesis and fabrication of -tricalciumphosphate (whitlockite) ceramics for potential prosthetic applications, *J. Mater. Sci.*, 14, 142, 1979.
71. **Karbe, E., Köster, K., Kramer, K., Heide, H., Kling, C., and König, R.,** Knochenwachstum in porösen Keramische Implantaten beim Hund, *Langenbecks Arch. Chir.*, 338, 109, 1975.
72. **Köster, K., Karbe, E., Kramer, H., Heide, H., and König, R.,** Experimentellen Knochenersatz durch resorbierbare calcium phosphat — Keramik, *Langenbecks Arch. Chir.*, 341, 77, 1976.
73. **Monroe, Z. A., Votawa, W., Bass., D. B., and McMullen, J.,** New calcium phosphate ceramic material for bone and tooth implants, *J. Dent. Res.*, 50, 860, 1971.
74. **Levin, M. P., Getter, L., Cutright, D. E., and Bhaskar, S. N.,** A comparison of iliac marrow and biodegradable ceramic in periodontal defects, *J. Biomed. Mater. Res.*, 9, 183, 1975.
75. **Cameron, H. U.,** Evaluation of a biodegradable ceramic, paper presented at the 25th Annual meeting of the ORS, San Francisco, February 20 to 22, 1979, 230.
76. **Peelen, J. G. J., Rejda, B. V., and de Groot, K.,** Preparation and properties of sintered hydroxylapatite, *Ceramurgia Int.*, 4, 71, 1978.
77. **White, E. W., Weber, J. N., Roy, D. M., Owen, E. L., Chiroff, R. T., and White, R. A.,** Replamineform porous biomaterials for hard tissue implant applications, *J. Biomed. Mater. Res. Symp.*, 6, 83, 1975.
78. **Chiroff, R. T., White, E. W., Weber, J. N., and Roy, D. M.,** Tissue ingrowth of replamineform implants, *J. Biomed. Mater. Res. Symp.*, 6, 29, 1975.
79. **Chiroff, R. T., White, R. A., White, E. W., Weber, J. N., and Roy, D. M.,** The restoration of articular surfaces overlying replamineform porous biomaterials, *J. Biomed. Mater. Res.*, 11, 165, 1977.
80. **Holmes, R. E.,** Bone regeneration within a coralline hydroxyapatite implant, *Plast. Reconstr. Surg.*, 63, 626, 1979.
81. **Evans, F. G.,** *Mechanical Properties of Bone*, Charles C Thomas, Springfield, Ill., 1973, chap. 10.
82. **Peyton, F. A.,** Materials in restorative dentistry, *Ann. N.Y. Acad. Sci.*, 146, 96, 1968.
83. **Bowen, R. L. and Rodriguez, M. S.,** Tensile strength and modulus of elasticity of tooth structure and several restorative materials, *J. Am. Dent. Assoc.*, 64, 378, 1962.
84. **Cameron, H. U., MacNab, I., and Pilliar, R. M.,** Evaluation of a biodegradable ceramic, *J. Biomed. Mater. Res.*, 11, 179, 1977.
85. **De With, G., Van Dijck, H. J. A., Hattu, N., and Prijs, K.,** Preparation, microstructure and mechanical properties of dense polycrystalline hydroxyapatite, *J. Mater. Sci.*, 16, 1592, 1961.
86. **Albee, F. H.,** Studies in bone growth — triple calciumphosphate as a stimulus to osteogenesis, *Ann. Surg.*, 71, 32, 1920.
87. **Haldeman, K. O. and Moore, J. M.,** Influence of a local excess of calcium and phosphorus on the healing of fractures, *Arch. Surg.*, 29, 385, 1934.
88. **Schram, W. R. and Fosdick, L. S.,** Studies in bone healing, *J. Oral Surg.*, 1, 191, 1943.

89. **Getter, L., Bhaskar, S. N., and Cutright, D. E.,** Three biodegradable calcium phosphate slurry implants in bone, *J. Oral Surg.*, 30, 263, 1972.

90. **Niwa, S., Sawai, K., Takahashi, S., Tagai, H., Ono, M., and Fukuda, Y.,** Experimental studies on the implantation of hydroxyapatite in the medullary canal of rabbits, Paper 4.10.4, presented at the 1st World Biomaterials Congress, Baden near Vienna, April 8, 1980.

91. **Gaberthüel, T. W. and Strub, J. R.,** Treatment of periodontal pockets with tricalciumphosphate in man: a preliminary report, *Schweiz. Monatsschr. Zahnheilkd.*, 87, 809, 1977.

92. **Köster, K., Heide, H., and König, R.,** Resorbierbare Calciumphosphatkeramik im Tierexperiment unter Belastung, *Langenbecks Arch. Chir.*, 343, 173, 1977.

93. **Howden, G. E.,** Tissue reaction to the bioceramic "Synthos", paper presented at the symposium Mechanical Properties of Biomaterials, Keele, September 13, 1978.

94. **Lemons, J. E. and Niemann, K. M. W.,** Porous tricalcium phosphate ceramic for bone replacement, paper presented at the 25th Annual Meeting of the ORS, San Francisco, February 20, 1979, 162.

95. **Lemons, J. E., Ballard, J. B., Culpepper, M. I., and Niemann, K. M. W.,** Porous tricalcium phosphate ceramic for segmental bone lesions, Paper 4.10.3, presented at the 1st World Biomaterials Congress, Baden near Vienna, April 8, 1980.

96. **Bhaskar, S. N., Brady, J. M., Getter, L., Gromer, M. F., and Driskell, T.,** Biodegradable ceramic implants in bone, *Oral Surg.*, 32, 336, 1971.

97. **Cutright, D. E., Bhasker, S. N., Brady, J. M., Getter, L., and Posey, W. R.,** Reaction of bone to tricalcium phosphate ceramic pellets, *Oral Surg. Oral Med. Oral Pathol.*, 33, 850, 1972.

98. **Mors, W. A. and Kaminski, E. J.,** Osteogenic replacement of tricalciumphosphate ceramic implants in the dog palate, *Arch. Oral. Biol.*, 20, 365, 1975.

99. **Rejda, B. V.,** Composite Materials for Hard Tissue Replacement, Ph.D. thesis, Free University of Amsterdam, Amsterdam, 1977.

100. **Rejda, B. V., Peelen, J. G. J., and de Groot, K.** Tricalciumphosphate as a bone substitute, *J. Bioeng.*, 1, 93, 1977.

101. **Rejda, B. V., Peelen, J. G. J., Vermeiden, J.H. M., and de Groot, K.,** Botvervangend materiaal vervaardigd uit calciumfosfaat, *Ned. Tijdschr. Geneeskd.*, 122, 625, 1978.

102. **Swart, J. G. N., Feenstra, L., Ponssen, H., and de Groot, K.,** Voorlopige klinische ervaringen met een botvervangingsmateriaal, gesinterd tricalciumphosphaat, *Ned. Tijdschr. Geneeskd.*, 123, 1421, 1979.

103. **Nery, E. B., Lynch, K. L., Hirthe, W. M. and Mueller, K. H.,** Bioceramic implants in surgically produced infrabony defects, *J. Periodontol.* 46, 328, 1975.

104. **Nery, E. B., Pflughoeft, F. A., Lynch, K. L., and Rooney, G. E.,** Functional loading of bioceramic augmented alveolar ridge — a pilot study, *J. Prosthet. Dent.*, 43, 338, 1980.

105. **Nery, E. B. and Lynch, K. L.,** Preliminary clinical studies of bioceramics in periodontal osseous defects, *J. Periodontol.*, 49, 523, 1978.

106. **Jarcho, M., Kay, J. F., Gumaer, K. I., Doremus, R. H., and Drobeck, H. P.,** Tissue, cellular and subcellular events at a bone-ceramic hydroxylapatite interface, *J. Bioeng.*, 1, 79, 1977.

107. **Kent, J., James, R., Finger, I., Jarcho, M., Taggert, J., and Cook, S.,** Augmentation of deficient edentulous alveolar ridges with dense polycrystalline hydroxylapatite: a preliminary report, Paper 3.8.2, presented at the 1st World Biomaterials Congress, Baden near Vienna, April 8, 1980.

108. **Salsbury, R. L., Quin, J. H., Gumaer, K. I., and Kent, J. W.,** Prevention of alveolar ridge resorption by placement of hydroxylapatite implants in fresh sockets, Trans. 7th Annu. Meeting of the Soc. for Biomaterials in conjunction with the 13th Int. Biomaterials Symposium, Troy, N.Y., 1981.

109. **Aoki, H., Kato, K., and Tabata, T.,** Osteocompatibility of apatite ceramics in mandibles, *Rep. Med. Dent. Eng.* 11, 35, 1977.

110. **Ogiso, M., Kaneda, H., Arasaki, J., Ishida, K., Shiota, M., Mitsuwa, T., Tabata, T., Yamazaki, Y., and Hidaks, T.,** Hydroxyapatite ceramic implants under occlusal function. Animal studies and clinical studies, Trans. 7th Annu. Meeting of the Soc. for Biomaterials in conjunction with the 13th Int. Biomaterials Symposium, Troy, N. Y., 1981.

111. **Denissen, H. W., Driessen, A., and Wolke, J. G. G.,** Dense apatite abutments, Trans. 7th Annu. Meeting of the Society for Biomaterials in conjunction with the 13th Int. Biomaterials Symposium, Troy, N.Y. 1981.

112. **Frame, J. W., Browne, R. M., and Brady, C. L.,** Hydroxyapatite as a bone substitute in the jaws, *Biomaterials*, 2, 19, 1981.

113. **Riess, G., Köster, K., and Reiner, R.,** Erste klinische und tierexperimentelle Erfahrungen mit Tricalcium phosphat (TCP)-Implantaten, *Dtsch. Zahnaertzl. Z.*, 33, 287, 1978.

114. **Riess, G. and Jacobs, H. G.,** Klinische Erfahrungen mit Tricalciumphosphat (TCP)-Implantaten, paper presented at Symposium: Der Heutigen Stand der Implantologie, Hamburg, December 8, 1979.

115. **Lavelle, C., Wedgwood, D., and Riess, G.,** A new implant philosophy, *J. Prosthet. Dent.*, 43, 71, 1980.

116. **Riess, G.,** Prinzipiën des TCP-Implantationssystems nach zweijähriger klinisher Erifahrung, *Dtsch. Zahnaertzl. Z.*, 35, 777, 1980.

117. **Ducheyne, P., Hench, L. L., Kagan, A., Martens, M., Burssens, A., and Mulier, J. C.,** The effect of hydroxyapatite impregnation on skeletal bonding of porous coated implants, *J. Biomed. Mater. Res.,* 14, 225, 1980.

118. **Mittelmeier, G. H., Harms, J., and Hanser, U.,** PMMA cement with carbon fiber reinforcement and apatite ingredients: mechanical properties and tissue reaction in animal tests, Poster 1.18, presented at the 1st World Biomaterials Congress, Baden near Vienna, April 8, 1980.

119. **Frame, J. W.,** Porous calcium sulphatedihydrate as a biodegradable implant in bone, *J. Dent.,* 3, 177, 1975.

120. **McGee, T. D. and Wood, J. L.,** Calcium-phosphate magnesium-aluminate osteoceramics, *J. Biomed. Mater. Res. Symp.,* 5, 137, 1974.

121. **Hentrich, R. L., Graves, G. A., Stein, M. G., and Bajpai, P. K.,** An evaluation of inert and resorbable ceramics for future clinical orthopedic applications, *J. Biomed. Mater. Res.,* 5, 25, 1971.

122. **Carvahlo, B. A., Bajpai, P. K., and Graves, G. A.,** Effect of resorbable calcium aluminate ceramics on regulation of calcium and phosphorous in rats, *Biomedicine,* 25, 130, 1976.

123. **Hammer, W. B., Topazian, R. G., McKinney, R. V., Jr., and Hulbert, S. F.,** Alveolar ridge augmentation with ceramics, *J. Dent. Res.,* 52, 356, 1973.

124. **Freeman, M. J., McCullem, D. E., Wolf, D., and Bajpai, P. K.,** Reconstruction of mandibular bone with alumino-calcium-phosphorus oxide ceramics, Trans. 7th Annu. Meeting of the Soc. for Biomaterials in conjunction with the 13th Int. Biomaterials Symposium, Troy, N.Y., 1981.

125. **Jarcho, M., Dombrowski, L. J., Salsbury, R. L., and Bondley, B. A.,** Fluoride uptake and dissolution behavior of a synthetic dental enamel-like substrate, *J. Dent. Res.,* 57, 917, 1978.

126. **Winter, M., Griss, P., de Groot, K., Tagai, H., Heimke, G., Van Dijck, H. J. A., and Sawai, K.,** Comparative histocompatibility testing of seven calciumphosphate ceramics, *Biomaterials,* 26, 159, 1981.

127. **Ferraro, J. W.,** Experimental evaluation of ceramic calcium phosphate as a substitute for bone grafts, *Plast. Reconstr. Surg.,* 63, 634, 1979.

128. **Kaiser, G., Wagner, W., Tetsch, P., and Köster, K.,** Zur Regeneration knöcherner Defects nach der Implantation resriboerbarer Calciumphosphat Keramik. Eine vergleichende klinische Untersuchung, *Dtsch. Zahnaertl. Z.,* 35, 108, 1980.

129. **McDavid, P. T., Boone, M. E., Kafrawy, A. H., and Mitchell, D. F.,** Effect of autogenous marrow and calcitonin on reactions to a ceramic. *J. Dent. Res.,* 58, 1478, 1979.

130. **Ducheyne, P. and de Groot, K.,** In vivo surface activity of a hydroxyapatite alveolar bone substitute, *J. Biomed. Mater. Res.,* 14, 225, 1980.

131. **Shima, T., Keller, J. T., Alvira, M. M., Mayfield, F. H., and Dunsker, S. B.,** Anterior cervical discectomy and interbody fusion: an experimental study using a synthetic tricalcium phosphate, *J. Neurosurg.,* 51, 533, 1979.

132. **Boyne, P. J., Fremming, B. D., Walsh, R., and Jarcho, M.,** Evaluation of a ceramic hydroxyapatite in femoral defects, *J. Dent. Res.,* 57a, 108, 1978.

133. **Köster, K., Ehard, H., Kubicek, J., and Heide, H.,** Experimentelle Anwendung von Kalziumphosphatgranulat zur Substitution von konventionelle Knochentransplantaten, *Z. Orthop.,* 118, 398, 1979.

134. **Osborn, J. F. and Newesely, H.,** Bonding osteogenesis induced by calcium phosphate implants, Paper 4.1.4, presented at the 1st World Biomaterials Congress, Baden near Vienna, April 8, 1980.

135. **Nancollas, G. H. and Wefel, J. S.,** Seeded growth of calcium phosphates: effect of different calcium phosphate seed material, *J. Dent. Res.,* 55, 617, 1976.

136. **De Putter, C., de Groot, K., and Sillevis Smit, P. A. E.,** Transmucosal implants of dense hydroxylapatite in dogs. Trans. of the 7th Annu. Meeting of the Society for Biomaterials in conjunction with the 13th Int. Biomaterials Symposium, Troy, N.Y., 1981.

137. **Cranin, A. N.,** Assessment of endosteal implants, Int. Congress of Implantology and Biomaterials in Stomatology, Kyoto, June 9, 1980.

INDEX

Y

Z

SRS
£58-45